Bulletin of the Academy of Pedagogic Sciences of the RSFSR
No. 106, 1959

Methods of Teaching Physics in Soviet Secondary Schools

Proceedings of the Institute of Teaching Methods

V. F. Yus'kovich, Editor

Translated from Russian by
A. Moscona

Israel Program for Scientific Translations
Jerusalem 1966

© 1966 Israel Program for Scientific Translations Ltd.

This book is a translation of
VOPROSY METODIKI OBUCHENIYA FIZIKE V SREDNEI SHKOLE
Trudy Instituta Metodov Obucheniya
Izdatel'stvo Akademii Pedagogicheskikh Nauk RSFSR
Moskva 1959

IPST Cat. No. 2111

Printed and Bound in Israel
by S. Monson, Jerusalem
and Wiener Bindery Ltd., Jerusalem

SOVIET PUBLISHER'S ANNOTATION

This book discusses the methodological aspects of teaching physics in secondary schools. The most topical amongst these are the present state of the method of physics, the development of the reasoning ability of students during the study of physics, and the experimental testing of physics teaching methods.

The volume is intended for scientific workers and secondary-school teachers.

TRANSLATOR'S FOREWORD

The problem of teaching methods in general, and of teaching science in particular, is a topical issue, to which educators are devoting an increasing amount of attention; this, coupled with recent Soviet achievements in the domain of physics, makes the present book a publication of specific interest.

The articles in this collection discuss the present Soviet approach to teaching physics, and offer suggestions for the formulation of a definite methodology, based not only on didactic principles and practical experience but also on ideological considerations; the articles reflect the aims of the Soviet school and the procedures by which they are expected to be achieved.

The reader may find the differences from the Western methods and approach no less enlightening than the similarities.

For this reason, as well as for the sake of completeness, no attempt has been made in the translation to eliminate some ideological digressions occurring in parts of the text, particularly in the second article of the collection ("The development of rational thought in the students during the teaching of physics in secondary school").

A Brief Description of the Soviet Secondary-School System
(given for better understanding of the articles)

Before the reform of 1958, two types of secondary schools existed in the Soviet Union: (a) the seven-year school (for the ages of seven to fourteen), which provided incomplete secondary education, after which the student could proceed to some vocational training; (b) the ten-year school (seven to seventeen) which provided a complete secondary education, and from which, upon graduation, the student could go on to higher studies. After the 1958 reform, the system was converted to an eight-year school (seven to fifteen), from which the students can go to the secondary school proper (three years — 9th to 11th grades), and thence to higher studies. Both before and after the reform, the secondary school included the first four grades of primary school (seven to eleven). Since the reform, more emphasis has been placed on "learning by doing" and on providing an all-round polytechnical knowledge.

CONTENTS

D. D. Galanin. The present situation in the method of high-school physics and its bearing on the polytechnical instruction system 1

V. F. Yus'kovich. The development of rational thought in students during the teaching of physics in secondary school 43

B. M. Yavorskii. "The electrical properties of the solid state" in school physics 89

E. E. Evenchik. Experimental findings in teaching "heat and work" in the 9th grade and "direct current" in the 10th grade 97

N. M. Shakhmaev. Some methodological aspects of teaching electromagnetic field phenomena in secondary school 106

S. E. Kamenetskii. The use of analogy in the secondary-school physics course 132

I. M. Rumyantsev. The training of students for practical work in physics 159

V. G. Razumovskii. Technical creativity of students in physics hobby groups 177

D. D. GALANIN

*THE PRESENT SITUATION IN THE METHOD OF HIGH-SCHOOL PHYSICS AND ITS BEARING ON THE POLYTECHNICAL INSTRUCTION SYSTEM**

"The verdict of reality is unequivocal — once that is
made known, it puts an end to controversy..."
Leonardo da Vinci

I. THE METHOD OF PHYSICS AS A SCIENTIFIC PEDAGOGICAL DISCIPLINE

The method of physics as a scientific discipline, like the method of any other subject taught in school, derives from two main sources.

The first source is the pedagogical sciences, and primarily didactics, which deals with the principles of teaching; the second source is the most remarkable of the sciences, physics, which has been blazing ever newer paths in the second half of the 20th century and has come to be the "model" of natural science.

At one time (the beginning of the 19th century), the model science was considered to be astronomy; nowadays, in the 20th century, astronomy is superseded by physics.

It should be added that the "correctness" of the laws of astronomy, the model science of the past, was contingent on the application of mathematics to the mechanical phenomena taking place in the universe. Now physics is an experimental science. Its method is founded on the empirical investigation of natural phenomena, in conjunction with precise reasoning and calculation.

In order to avoid many misunderstandings in the sequel, it should be also stated that the method of physics (as that of any other subject), when developed as a scientific pedagogical discipline, guarantees, if its conclusions are followed up, the effective teaching of the subject in school. This pedagogical science cannot, however, neatly define each step, and the whole procedure of teaching. Teaching, being an activity in which the will and consciousness of the teacher cooperate with those of the students, is a process which does not lend itself to a precisely defined regulation. The emotional factor plays an important part in teaching, and the latter may therefore be termed an art rather than an activity laid along strictly scientific lines.

The advice of veteran teachers to young ones usually refers not so much to the "theoretical method", so to speak, as to the art of teaching and to

---

* The present paper is based on a report presented by the author at an enlarged session of the Laboratory of the Methodology of Physics of the Institute of Teaching Methods of the Academy of Pedagogic Sciences, RSFSR, on 23 April 1956. The text was subsequently revised and expanded.

those subjective aspects of the pedagogical process which are not amenable to scientific formulation.

This state of affairs as regards the methodological procedure of teaching is due partly to an insufficiently developed method of physics. But this situation also reflects the inherent feature of methodology as a scientific discipline.

It should be added that in the development of any science there are two periods: the formative period with the discovery of fundamental laws and methods, and the period of practical application of the established scientific propositions. Scientific propositions which, to begin with, appear completely abstract and irrelevant to actual practice prove in due course of time to be the basis of extensive and fruitful application.

The history of science provides examples of how detrimental it could be to science to ignore theory or rather abstract and general principles, which constitute the basis of the scientific method.

An excessive practical slant proves detrimental to the development of any scientific discipline.

However, the development of science may be inhibited in the same measure by a dissociation from practice. "Do not forget," as Leonardo da Vinci says, "to adduce to each proposition its practical application, lest your science be useless." The scientific method of physics is undergoing at present a certain crisis, a period of accumulation of material, and therefore charges are often voiced against it that it fails to be of assistance in direct teaching. We hope to show in our survey that this is not true, and that the development of the method of physics has been, is, and will be of tremendous help in teaching.

Let us return to the description of the main sources of methodological synthesis. The current stage of development of general pedagogical sciences is inadequate, despite the fact that they have been in existence for several centuries. The method of physics as an individual specialized discipline cannot, unfortunately, rest fully on the objective, clear-cut findings of the pedagogical sciences. In particular, specific difficulties arise owing to the lack in pedagogy of a well worked out "science of secondary education." The Leningrad Institute of Pedagogy of the APN RSFSR* is just beginning to tackle the problem. The development of the science of secondary education in the Soviet Union will doubtless introduce into all the individual methods, and among them the method of physics, a number of new ideas and fresh objectives. For the time being, until this scientific research has been done, the problems and their solution may be stated only as hypotheses or as tradition. The definition of the objective of secondary education is to a large extent a political problem. It is fundamentally to be solved by the directives of the party and the government. The pedagogical sciences help, on the one hand, in working out the party and government directives by studying the secondary educational system and contemplating its relations with its other aspects of social life; on the other hand, the pedagogical sciences ought to assist the authorities in putting these directives into effect.

After the resolutions of the 19th and 20th Congresses of the Communist Party on polytechnical training, and especially after the adoption of the law on the consolidation of the link between school and life and on the further

---

* [Akademiya Pedagogicheskikh Nauk Rossiskoi Sovetskoi Federativnoi Sotsialisticheskoi Respubliki — Academy of Pedagogic Sciences of the Russian Soviet Federative Socialist Republic.]

development of the system of public education in the USSR," new tasks and problems have been set before pedagogical science.

In connection with the new tasks set to the national economy in the Soviet Union, as established by the 21st Congress of the Communist Party, the heretofore prevailing conception on secondary education must be considerably modified and expanded. In this respect pedagogical science lags significantly behind life, and this hinders the development of the method of all school subjects, and particularly the method of physics.

Another general pedagogical discipline whose poor development considerably inhibits the method of physics is general didactics.

Since the time of Jan Amos Komensky, 300 years back, didactics developed the principles of primary education, which applied itself to reading, writing, and arithmetic. Now even in the 20th century, didactics has far from worked out adequate principles of teaching on a higher level, a complete secondary education. The individual methodological disciplines are therefore often obliged to proceed on their own to a large measure, drawing upon the teaching experience in their respective fields alone.

It is perfectly clear that the teaching process depends to a very large extent on the subject matter which is being studied; it is possible, though, to find in the various subjects material of similar nature and to indicate the best ways of teaching it. These general propositions have to be worked out by didactics, on the basis of teaching experience in the various subjects. Didactics might well profit thereby as well, similarly to the way fruit trees are improved by cross-pollination.

We pause to consider another important factor which should be given special attention when setting up methodology as a scientific discipline in the conditions of the Soviet Union. We mean the mass scale on which physics is taught, which did not exist in Russia before the Revolution, when physics teachers were for some decades supplied only by universities, where the method of physics was not taught at all.

The pressing need for developing methodology as a special pedagogical subject is dictated by the hectic development of education produced in any country after a proletarian social revolution, which inevitably entails a cultural revolution as well.

Soon after the October Revolution it became necessary to evolve promptly methods for the training of physics teachers. This will be discussed in more detail later on, when we present the history of Soviet physics methodology.

Let us now turn to the second source of the method, to the taught subject itself, that is to say, physics. The method of physics differs from that of any other subject due to the following features of physics as a science:

1. Physics is an experimental science. It draws its subject matter from the empirical investigation of natural phenomena, relating to these phenomena actively, by seeking the means of reproducing and studying them through experiment.

2. Physics deals with the elementary forms of motion of matter (it should be borne in mind that these terms are used in the philosophical sense). Physical science studies the simplest and most general properties of nature, those which are most prevalent and constitute the basis of more involved phenomena than are studied by the other natural sciences. The general properties of any kind of matter, molecules, electrical elementary

particles, and electric, magnetic and electromagnetic fields, always "fall back upon physics", whenever the matter or fields might be encountered, be it in a human brain cell, at the interior of stars, or in any machine. Therein resides the universality of the phenomena studied by physics and their vast scope, and hence their importance in general education.

3. Since its very inception, physics has concerned itself with more than just the recording and description of phenomena or just their classification (as, for example, was the case with botany for a long time). Physics aims at the essence of any phenomena, draws up general laws and casts them into quantitative form. A case in point is one of the oldest physical laws, Archimedes' principle. This law is interesting, because it establishes a quantitative relationship, it is provable by general considerations on the equilibrium of objects in liquids, and because it is of extensive practical use and is verified thereby.

A distinctive characteristic of physics, which has been clearly apparent throughout its history, is that it aims at eliciting the intrinsic nature of phenomena and deriving from them general principles, which embrace a wide range of empirical facts and provide the means of predicting new, experimentally still undiscovered ones. In this respect suffice it to recall the laws of dynamics and Maxwell's equations, or the molecular kinetic theory and wave mechanics. These general principles are ultimately given a rigorous quantitative formulation, and that links physics with applied mathematics. This feature makes the teaching of physics a very interesting, though often difficult task which imposes on its method its own specific demands. In the prevailing methods of physics this particular aspect is not given due consideration.

4. Lastly, another characteristic of physics stemming from the ubiquity of its subject matter ("the elementary forms of motion of matter") is its direct bearing on technology. The discoveries of physics have in many instances given rise to special technical and applied sciences, necessary to the practical man such as the engineer and technologist or the designer and constructor. Newton, for instance, made use of physical methods of investigation to construct his theoretical, or physical mechanics which soon provided a fruitful basis for applied or structural mechanics, which is one of the main subjects in any vocational education. The physical conception about the elastic properties of solids gave rise to the technological discipline "strength of materials". This discipline subsequently branched off into another specialized engineering discipline, referred to as "machine parts". Electrical engineering, followed by radio, heat, and lighting engineering, the optical and vacuum technologies, and, finally, nuclear power engineering have all evolved from physics.

On the other hand, the rapidly developing technology and industry pose new and difficult problems for the physicist.

The discovery of physical laws covering a broad range of natural phenomena is thus pertinent not only to the investigation of these phenomena but also to their application for the improvement of the national economy.

This practical import of each new advance in physics, such as the application of electricity or radioactive isotopes, was not taken into account (or very little so) in the pre-revolutionary methodology of physics. Due to the conservatism of pedagogy, physics lagged behind the development of science and turned into "notebook" physics, a kind of "mathematized physics". If

Soviet physics methodology is to be developed as a scientific pedagogical discipline, this situation cannot be tolerated. It often was, however.

We have made it a point to discuss the characteristic aspects of physics as a science, because they set the pattern for its method. It appears to us that the method of physics has not given due consideration to these aspects, of the very science it is supposed to learn how to teach.

In conclusion to these preliminary remarks let us examine the general problem of "science instruction", which is intimately connected with the further development of science itself.

The "original methodologist," who lays the foundation to all subsequent pedagogical undertakings, is the scientist himself, when he first publishes his discovery.

The researcher who then carries on no longer repeats the original work, unless he can substantially add to it. It often happens, though, that earlier works are explained by an eminent scientist, who creates new concepts which are used as a basis for further development. His exposition then assumes a pedagogical, methodological role.

It also happens quite often, especially in a science as developed as modern physics, that a scientist is called upon to produce "surveys" of a compilatory nature which should enable young scientists readily to acquaint themselves with scientific developments. Surveys of this kind, in the form of comprehensive articles and sometimes books, are proliferating at an ever increasing rate in the twentieth century. The output of the world-wide army of physicists is becoming so large that a young scientist finds himself at a loss when he has to cope with previous researches before tackling any original work of his own.

In this process of "scientific information" (for which there is in the USSR Academy of Sciences a whole "Institute of Scientific Information") the pedagogical aspect detaches itself in a more significant and definite manner, finally coalescing into the method of teaching university physics and, ultimately, into the method of teaching high-school physics.

This is where the pedagogical teaching procedure takes its point of departure. In this way "textbooks" are produced in the various sciences.

The way in which presently current methods of exposition have been evolved, i.e., the process of formation of a "student physics" out of the "physics for scientists" has not yet been adequately studied from the historical standpoint. If the problems involved in the method of physics are contemplated from a wider angle, the study of this process could be included within its scope and would probably prove to be of considerable advantage.

The growth of any science is promoted in direct measure to the ease with which its principles can be assimilated by the budding specialists in the rising generation.

Every scientist is constantly faced with the need of carefully considering the clearest and most intelligible way in which to present his findings, no matter how far he may be removed from teaching. This same problem of the most intelligible way of presentation constitutes the object of the method of physics, and is intrinsically completely analogous to the one which daily confronts the physics teacher.

Among the famous physicists of modern times, two have gained recognition as eminent pedagogues — the French physicist Paul Langevin and the Italian physicist Enrico Fermi, the founder of nuclear science. Among the Russian physicists of the past, we have the notable physicists O. D. Khvol'son and N. A. Umov. In the works of presently living physicists deserving special attention are those of Academician A. F. Yoffe; to him are due a number of textbooks and some popular-science books. Mention should also be made of the valuable attempt to produce a "Primer of Physics" made by a group of scientific workers of the Academy of Sciences of the USSR, headed by G. S. Landsberg.

We feel that the scientific pedagogical aspects of methodology will repay thoughtful consideration. It is a pity, though, that our prominent scientists all too rarely devote themselves to methodological problems.

II. A BRIEF HISTORY OF PHYSICS METHODOLOGY IN THE SOVIET UNION AND AN ACCOUNT OF ITS BASIC REFERENCE WORKS

1. The development of the method of physics in the Soviet Union was greatly enhanced by the tremendous expansion of the secondary-school network, which required the prompt training of a large number of physics teachers. The teaching of physics in secondary schools assumed such proportions that physics teachers could no longer be left to develop their capacity according to personal taste, as was the case in Tsarist Russia. The state was confronted with the pressing need of training many hundreds and even thousands of physics teachers in pedagogical institutes. One of the main subjects in the curriculum of these institutes had to be the method of physics. It was therefore necessary to unify the diversity of prevailing views on the best way of teaching physics. This rapid development of the method of physics took place in the beginning of the thirties.

The Soviet methodology of physics had at that time a basis from which to proceed, of course; it could draw upon the rich fund of information left by the works of Professor Shvedov of Odessa, Assistant Professor V. V. Lermantov of Leningrad, and the methodological papers of Professor N. V. Kashin, which were produced while training teachers in the Shelaputin Institute prior to the Revolution. There was little general teaching experience in physics at the time. It was necessary to work out and put into practice without delay teaching procedures, notwithstanding the fact that the qualifications of the available personnel were below par. This placed a special responsibility on the authors of methodological books.

The academic research institutions of the time (The Central Physics Pedagogical Institute founded by A. V. Tsinger, at the Polytechnical Museum in Moscow and the Institute of Scientific Pedagogy in the former "Solyanoi Gorodok"* in Leningrad) could not be readily reorganized, being "so close to the trees as not to see the woods". The staff of these institutes was often constrained to seek ways of studying physics only in as far as the improvement of the efficiency of labor was concerned.

The methodological investigations of the first years after the Revolution may still prove of value to authors of courses on the method of physics.

Let us mention some of the books which appeared in those years, and which played a significant role in the consolidation of the method.

* [An urban district in Leningrad.]

In 1922 a second edition of N. V. Kashin's "Methodology of Physics" (Metodika fiziki) and I. Glinka's "Physics Methods in Practice" (Opyt po metodike fiziki) appeared, and in 1923 V. V. Lermantov's "Methodology of Physics" made its appearance. In 1924 the "Collected Papers of Physico-Mathematical Topics and the Ways of Teaching Them" (Sbornik statei po voprosam fiziko-matematicheskikh nauk i ikh prepodavaniya) was published, in which the physics editor was A. I. Bachinskii; this collection served as a basis for a series of periodicals. In 1925 a Russian translation of the book "The Teaching of Physics in Secondary Education" by the noted American methodologist Mann was published. This textbook, written in conjunction with Twiss, was translated before the Revolution and was later reprinted several times.

Notwithstanding the lucidity of exposition, Mann's little book did not do much in the way of establishing Soviet methodology.

As may be seen from this list, an attempt was made to publish all available material without paying particular attention to the unity of the views expressed. It suffices to compare N. V. Kashin's methodology, which displayed a pronounced French influence, with the almost "mischievous", purely Russian statement of V. V. Lermantov's. Out of this literature, only Kashin's course is in any measure adapted to teaching purposes. Kashin's conception of teaching physics, failed, however, to meet the demands of physics studies in vocational schools, where the curriculum was based on the "unit" teaching system.*

In 1927 the collection "Physics, Chemistry, Mathematics, and Engineering in the Vocational School" (Fizika, khimiya, matematika, tekhnika v trudovoi shkole), edited by P. P. Lebedev (Professor of Chemistry) appeared, and became the starting point of many collections, and later on of a methodology journal. In 1929 Professor G. G. De Metz published his "General Methodology of Teaching Physics" (Obshchaya metodika prepodavaniya fiziki), which was the first course on the method of physics to take into account the specific requirements of vocational schools.

At the beginning of the thirties the number of books on the method of physics had already considerably grown. However, the number of published copies was by far inadequate to cover the multiplying needs of teachers.

At that time the basic issues of teaching physics as a special subject had not yet been settled, but much had been done which even today, thirty odd years later, may still be of value. Let us cite the main achievements of that period:

1) "Notebook physics" was subjected to thorough ideological criticism, and it was firmly established that the course in physics should conform to the needs of real life, and thus to the requirements of vocational schools.

2) During those years many home-made devices were designed and built and various methods were developed for the demonstration of physical experiments without the help of special instruments. The practical experience gained in this do-it-yourself activity has not been sufficiently studied and generalized to this day. It has been to a considerable extent irretrievably lost, since it had not been published in print. The authors of these home-made devices later described whatever they still happened to remember, and a few specimens remained in the physics laboratories of some schools.

---

* [The "unit" teaching system was prevalent in the Soviet Union between the early twenties and the beginning of the thirties. The system consisted of selected "unit" topics of study, each topic incorporating various pertinent sections of the material from the different subjects on the school curriculum (such as language, science, mathematics). This precluded a consistent treatment and continuity in the study of the individual school subjects, resulting in fragmented and uncoordinated knowledge, and the system was eventually abandoned, in 1931.]

3) The existing environment of the school-goer was then also carefully examined so as to select equipment suitable for the instruction of physics. In those years Russia was a poor and agrarian country and did not have the products of advanced technology by which the student is surrounded today. It was necessary to content oneself with such items as, for instance, a well (F. N. Krasikov), a samovar, and some hand-made instruments. In the conditions of that time, the wiring of the lighting system and the connection of an old lamp socket or of an electrical plug by means of a piece of used wire appeared as the height of industrial achievement, and the replacement of a burnt-out fuse with a clip of wire bespoke an impressive proficiency in electrical engineering.

2. During the subsequent period, extending over the early thirties, the development of methodology did not proceed too smoothly. This was the period of production-plant apprenticeship schools, a period of "technology-physics." A large number of various booklets were printed at the time and periodicals of methodology started appearing. These soft-cover publications, which came out in lots of several thousand copies, could not meet the overall school requirements but they provided particularly successful educators with an outlet for commenting on their experience. The thirties are noted for the establishment of the five-year plans, whose effective execution has made the Soviet Union into a great industrial power.

The technological progress which the pedagogues of the twenties envisioned to be embodied in the study of physics was in fact realized in everyday life. The universal pursuit of knowledge in which the masses engaged outstripped the school instruction. The school lagged behind even in its basic task of teaching reading, writing, and arithmetic, not to speak of such subjects as physics, which require special facilities. In those years all the members of the population, young and old alike, were studying. This general educational drive was mainly directed towards technical subjects. At the same time, Lenin's electrification plan was put into practice, amateur radio developed at a rapid pace, and a large number of automobiles and tractors — still imported at the time — made their appearance, with a concurrent demand for drivers and mechanics.

A large number of popular science journals started appearing, but their effect in raising the general level of knowledge in physics has not yet been properly assessed. At that time physics was still poorly taught in school, but knowledge in physics and closely related technical topics was rapidly disseminated.

Now what were the basic views on methodology at the time ? This may be made apparent from the following quotation. An introductory note to the curriculum of 1927 states that "Physics ... is faced with considerable difficulties. The main difficulty is that the problems raised by the "unit" teaching curriculum require a fairly broad basis of technological information... The second difficulty is that the subject matter in physics may be reclassified within the curriculum only up to a certain point, if the course is to be economically constituted and methodologically well-founded."*

In fact, it is perfectly clear that given a curriculum of "unit" teaching there is no way of constituting a "methodologically well-founded" course.

As we have already mentioned, the first Soviet methodology of physics, due to Professor De Metz of Kiev University, made its appearance in 1929, as an attempt to provide a solution to the problem. In one part of the methodology an attempt

---

*I.I.Sokolov, Metodika fiziki (Methodology of Physics), p.19. 1934.

is made to work out a theoretical basis for a possible nonsystematic way of teaching physics. However, the esteemed author, to whom Russian physics teachers owe the benefit of the journal "Physical Review" (Fizicheskoe obozrenie) of which he was the second editor (the first was Professor Zilov), devoted more space to presenting the way physics was taught abroad than to trying to find and analyze new means of teaching physics under the new conditions.

Beginning with the thirties, physics became an independent subject in the curriculum. From that moment, it became necessary to train teachers at the pedagogical institutes in the method of physics, and in the science and art of teaching. It may well have been the first time in man's history that a teachers' training program was instituted on such a grand scale.

This "pedagogical necessity" posed the problem of creating a comprehenisve course in the method of physics, instead of papers on individual topics. The subject of study of aspiring teachers had to be the teaching of physics as a whole. Lecturers and methodologists were thus confronted with the difficult task of generalizing and supplementing, often at short notice, the material that had not been properly worked out.

The first methodology courses for this purpose — I. I. Sokolov's in Moscow, and the one edited by P. A. Znamenskii in Leningrad, compiled with the help of the noted methodologists E. N. Kel'zi and I. A. Chelyustkin — fluctuated considerably in quality and did not discuss equally thoroughly all the points of the method of physics. And yet this was one of the things that pedagogy students and new teachers most acutely needed.

From our historical survey, we pass on to a systematic review of the subject, since the basic methodological viewpoint expressed in the methodology courses are of topical importance for the present as well.

3. Out of the three complete courses on the methodology of physics currently available, those of P. A. Znamenskii, of I. I. Sokolov, and of E. N. Goryachkin, the one of greatest objective significance, which endeavors to achieve a scientific synthesis, is the methodology course of P. A. Znamenskii (2nd edition, Uchpedgiz, 1954).

If one were to describe concisely the general tenor of this methodology, it may be said to have drawn largely upon the traditions laid by O. D. Khvol'son.

The author of this book, P. A. Znamenskii, was an active participant in all the conferences and commissions before and after the Revolution, which dealt with physics methodology problems, and the first professor in the method of physics at the Gertsen Pedagogical Institute. To a certain extent Znamenskii is a successor or even a pupil of N. S. Drentel'n, B. Yu. Kol'be, V. V. Lermantov, A. P. Afanas'ev, and many other Leningrad physicists, scholars, and teachers, who never failed to maintain a high standard of science and culture.

Znamenskii's methodology is the direct sequel of two methods of physics, the one formulated by Znamenskii, E. N. Kel'zi, and I. A. Chelyustkin, and the other by Znamenskii, V. N. Ziber, E. N. Kel'zi, and M. Yu. Piotrovskii (10th-grade course), formulated in the 1934-35 school year.

In his method Znamenskii has tried to find a generalized solution to methodological problems by analyzing and synthesizing various opinions. The author followed closely the activity of Leningrad pedagogues and gave expression to their views in his work. Thus the particular solutions which he provides fit nicely within the general structure of the method of physics.

Out of the three above-mentioned methodology courses, Znamenskii's proves to be the best suited to the purposes of teaching, mainly because of its conciseness. Any redundancy has been removed from it, and the resultant material is set out in a soberly thought-out system. The general impression derived from the book, which is enhanced as one goes along, is one of restraint. In all fairness it should be said that the author carefully appraises any new ideas or methodological formulations appearing in articles, theses, and reports, so as to incorporate them into his methodological system. Thus the second edition of the book covers A. A. Pokrovskii's work on physical experimentation and D. D. Galanin's work on the method of mechanics.

An interesting instance of the way of exposition in Znamenskii's methodology of physics is provided by Chapter III, "Polytechnical training in the study of physics" (Politekhnicheskoe obuchenie v uchebnom protsesse po fizike), pp. 36-43 of the second edition, in which the new objectives set before the school are stated quite briefly but clearly. For instance, on p. 37 the author writes: "The introduction of polytechnical training denotes a new, big step towards achieving a well-developed educational procedure in secondary school. The realization of polytechnical training will bring about substantial changes in the content and method of school studies, and will impart a new aspect to the school."

One can agree to this readily enough, but it must be said that Znamenskii himself falls short of fulfilling these provisions in his methodology. The specialized part of Znamenskii's methodology rests largely on the old attitudes as to the proper, scientifically correct ways of studying the fundamentals of physics.

A deficiency of Znamenskii's methodology stems from the fact that it does not deal sufficiently with the teaching of physics in high school. Quite to the contrary, some of the suggestions made in the specialized part raise the teaching level beyond curriculum requirements (e. g., the undular nature of light, fluid mechanics). The author makes no attempt to find new and simpler means of presentation, which could help the students in studying these topics, but rather sticks to the "classical", so to speak, traditional methods of presentation, and works them up to a considerable degree of sophistication.

The method of teaching physics as presented by Znamenskii is designed for the capable student living in a big industrial center and studying in a school with good facilities. The reader of the methodology will find no suggestion as to how to proceed when such conditions do not exist. Neither will he find in the Znamenskii methodology a historical account of how any particular methodological proposition came to be formulated.

4. The second methodology course, which appeared the first (in 1934) is that of I. I. Sokolov; this course is quite different from Znamenskii's.

I. I. Sokolov, whose eightieth anniversary was recently celebrated, finished the Moscow University, where he was the pupil of A. G. Stoletov and N. A. Umov. Like Znamenskii, Sokolov also participated in the conferences and meetings on the teaching of physics before and after the Revolution. The ideal that Sokolov strove to achieve was a rigorously finished, polished version of Stoletov's exposition.

The methodology, especially in its first edition of 1934, is uneven in its presentation. There is no indication that the author took into consideration the possibilities of the students and restricted himself to the presentation of particular details. Quite often the author fails to take into consideration his pedagogical objectives and elaborates the subject in great detail, to the extent of including observations of a polemic nature. There is a historical slant in the treatment of various questions, explaining the way in which the particular methods of exposition were developed. This is quite a valid method of presentation for a scientific treatise, which the first edition of Sokolov's methodology actually turned out to be. In a book intended for teaching purposes, this kind of exposition is bound to confuse the pedagogy students, who cannot tell where the discussion of previous stages ends and the statement of definitive conclusions begins. This particular feature, however, makes Sokolov's methodology highly valuable for the methodologist or the experienced teacher, as the book can provide them with considerable food for thought. One may find oneself disagreeing with Sokolov on various points, but one will still benefit considerably from what the author has to say. There are appreciable differences between the first edition of 1934 (the Foreword is signed 15 May, 1933), the second edition of 1936, and the third edition of 1951. This indicates that the author has made constant efforts to improve his book.

In the first edition considerable space has been devoted to the systematic treatment of the course in physics and to the teaching of physics in workers' faculties and technical schools. In the third edition, Chapter V — "A brief history of the Soviet physics curriculum in secondary schools" and Chapter VI — "Principles for the formulation of the Soviet curriculum in physics" have be n retained, but any controversial statements have been eliminated from the discussion. The last chapter analyzes various aspects of holding a physics course in secondary schools.

The second part of the methodology (Chapters VII—XV) deals with teaching methods and the organization of the pedagogical procedure. Notwithstanding the fact that the book contains a special chapter on "Physics and technology" (pp. 26 -29 — no more than 3 pages!) and refers to the importance of technology and to the desirability of being acquainted with the problems of production for polytechnical education, this aspect of teaching physics is the weakest point in Sokolov's methodology.

Sokolov correctly writes on p. 36 (third edition) that "the physics teacher must not confine himself to imparting to the students a definite amount of facts." However, we often fail to find in the specialized part of the methodology any suggestions as to what exactly this amount of facts is to be supplemented with in order to achieve the appropriate study of physics. The specialized part gives a rather poor exposition of modern views on many contexts; thus the modern physics theories presented in Sokolov's methodology are not up to providing a basis on which to constitute the main branches of physics for a secondary-school course.

Despite the above deficiency, Sokolov's exposition in the specialized part gives the teacher more than does Znamenskii's, which is much too brief. This methodology has, accordingly, its advantages, though it is by and large not as suitable for pedagogy students as Znamenskii's methodology.

Let us note one more fact which complicates the exposition in the specialized part of Sokolov's methodology.

This complication is associated with the fact that often the m e t h o d of any particular topic is replaced by a factual account of the topic itself.

This is due to the fact that, as observation has shown, both pedagogy students and young teachers have a poor knowledge of elementary physics.

It may be said that there are very few reference books from which this knowledge could be had. A teacher is expected to have extensive but detailed knowledge of elementary physics, while standard secondary-school textbooks provide a limited amount of material; the authors of university textbooks, on the other hand, assume that these elementary notions are already known.

This has caused the need of providing not only a methodological exposition but also a factual account of the various topics in elementary physics.

It is clearly indispensable to have a printed course in elementary physics giving a fairly wide coverage of the history of physics and the exposition of modern use. Such a course should obviously also contain information on conducting physical experiments.

This "encyclopaedia of elementary physics" should provide the pedagogy student and the beginning teacher with comprehensive material on any subject likely to arise in the course of teaching.

This requirement was formerly fulfilled to a certain extent by the first edition of Grimsel's Physics Course; however, in the subsequent revisions this particular feature of the course disappeared. O. D. Khvol'son's "Kurs fiziki" (A Physics Course) and the "Fizicheskii slovar'" (A Physical Dictionary), edited by P. N. Belikov, could be of some aid, but both these publications are intended for the scientific worker and not for the secondary-school physics teacher and have become rather old-fashioned.

There is a definite need for the publication of such a physics course, but its compilation involves considerable effort. The principal advantage of the course must be simplicity of exposition, coupled with a broad range of examples covering the technological applications of physics. In spite of its being elementary, the exposition should reflect modern views on each topic.*

5. The third extensive work in four volumes on the methodology of physics, which covers only the elementary physics course, i.e., for the 6th and 7th grades, is due to E. N. Goryachkin.

The characteristic feature of Goryachkin's methodology is that it is primarily designed for the young teacher rather than for the student of a pedagogical institute. The author appears to have kept in view not the students but the young teachers who have just finished a pedagogical institute. He visualizes their work in school and tries to provide helpful hints. The author is very familiar with the needs of young teachers, as he worked for many years at a well-known pedagogical institution, the one-time Karl Liebknecht Institute in Moscow. In his main objectives (which are possibly not fully implemented throughout this substantial work), Goryachkin places more emphasis than other authors on the presentation of modern physics as a science and its connection with technology in the process of teaching. Physics as taught by Goryachkin's method is, in spite of its elementary level, appreciably closer to life and to modern scientific views than physics as taught by Solokov's method. In particular, it must be noted that physics as taught by Goryachkin's method is closely related to

---

* The "Elementarnyi uchebnik fiziki" (A Physics Primer) appearing in the third edition under the editorship of G. S. Landsberg meets this requirement in part, but it is actually designed for the student and does not provide the details necessary to the teacher.

technology. Technology, in Goryachkin's method, is organically included in the presentation of each topic, so that no discontinuity occurs between science and technology, as is the case with some authors.

We adduce some of the basic propositions in Goryachkin's physics methodology (pp. 10 - 11).

"It is characteristic of our Soviet school that there is and can be no contradiction between the content of science and that of the school subject..."

"... It thus follows that the school physics course must be so constructed as to provide a well-defined system of physical knowledge and produce a scientific outlook."

"The exposition will be scientific if it is constructed in accordance with leading scientific views..."

"... The pedagogical exposition cannot be merely a "compilation" or just a "simplification"; it must present a consistent set of viewpoints and display the distinctive characteristic of scientific creativity."

This methodological standpoint does not "look back" into "the past", but impels the teaching of physics forward, setting as its ideal the very way of teaching physics which is in the process of being organized at present in the Soviet school and whose effectiveness will be increased with improved material conditions. These basic objectives define the problems of teaching physics, the content of the physics course and the methods of implementing it in the process of teaching.

A prominent feature in Goryachkin's methodology is the "pedagogical sovereignty" of experiment. In point of fact, in the elementary teaching of physics, the content of the physics course, and the methods of implementing it in the process of teaching, experiment is a basic didactic procedure; the other teaching aids, i.e., comments by the teacher and in the textbook, illustrations in books, on the blackboard, and in the students notebooks, are of course necessary to enable the students to grasp the basic material with which they are presented by physical experiment. Goryachkin's methodology thus dwells at length on the part played by experiment in teaching practice, and on the psychology and logic of this participation.

This basic requirement determines the subject matter of the following volumes in the methodology: the second volume presents experimental methods and techniques classified according to topics, and describes the required physical equipment; the third volume deals with home-made simplified devices, as it is not always possible to count on there being at school a full range of factory-made instruments, and, according to the methodology, experimentation is an indispensable part of the teaching of physics. The writing of the third volume was also conditioned by the author's desire to provide comprehensive material for the practical poly-technical training of the students (even though the book appeared before the law on polytechnical training was passed). The fourth volume describes the drawings and illustrations to be used in the physics classes, considering them as a particular kind of language by means of which many physical facts can be imparted.

The perusal of the main headings of this interesting methodology suffices to show that it constitutes a completely new methodological construction.

Mention should be also made of a particular feature of the specialized section in Goryachkin's methodology, which deals with the individual topics in the curriculum.

The first volume brings into consideration the content and method of the introductory discussion of each topic, and also presents historical material, laboratory work, visual aids, studies outside the classroom, and home assignments.

Each section in the main, first volume dealing with any individual subject contains references to the corresponding section in Volume 2, where experimental techniques are described in detail, in Volume 3, where the appropriate simplified instruments are described, and in Volume 4, where it is explained how to make the drawings necessary for teaching the subject.

This method of presentation is, of course, not very well adapted to the studies of the methodology student preparing for any examination on the method of physics. It does, however, provide a good background for the young teachers, who will find it a helpful reference work. The author has received more than one thousand (!) letters of appreciation from methodologists of physics and from teachers. Here is a representative sample: "I have never yet come across such a complete manual-encyclopaedia on the method of physics either in our or the foreign literature, and I have read more than a few."

The first volume of Goryachkin's methodology has been translated into German and adopted as a textbook for the training of teachers in the German Democratic Republic.

It is a much more difficult matter to put into practice on a consistently high level such a far-reaching project than it is to conceive it. This voluminous work thus suffers from some inadequacies, but the tremendous public response that it has elicited indicates that Goryachkin succeeded in making a new significant step in the development of the Soviet method of physics. This is the first time that the physics teacher has at his disposal such a comprehensive methodology course.

A drawback of Goryachkin's methodology that is often mentioned is the fact that the level of elementary physics is set much too high and that insufficient attention is paid to the ways in which the acquisition of knowledge may be facilitated.

It is possible, however, that the polytechnical training instituted at present in the primary classes could greatly enhance the ability of students to assimilate the material in physics.

It may be assumed that the full implementation of the pedagogical methods recommended in Goryachkin's methodology should make the learning of physics much easier. Now, the directions given in the methodology have never been fully applied in teaching practice. In fact, the validity of the teaching methods formulated in the various methodologies still have to be put to a practical test.

It is indubitable that the thorough application of these methods will place the students' knowledge on a firmer basis thus ensuring that a wider field of knowledge could be more easily imparted than could limited material but with poor teaching techniques.

However, this divergence between theory and practice occurs fairly commonly. It is up to the educational authorities to correct the situation.

6. In 1952 the "Essays on Physics Teaching Methods" (Ocherki po metodike prepodavaniya fiziki) (Volume 1) of the two eminent Ukrainian methodologists O. K. Babenko and M. U. Rozenberg appeared, in Ukrainian. The essays deal with the 8th to 10th grades only; this is possibly due to the fact

that Goryachkin's methodology for the 6th—7th grades has been translated into Ukrainian.

As far as may be judged by the first volume, the methodology exhibits a strong practical slant and is devoted to specialized aspects. The book describes at length experimental techniques, often in an original way. This methodology constitutes quite a useful reference book. However, a "new" methodology for the 8th—10th grades, discussing the teaching of physics under the polytechnical training program, still remains to be written. The Physics Laboratory of the Institute of Teaching Methods of the Academy of Pedagogical Sciences of the RSFSR is at present making an attempt to produce such a methodology.

This completes our general, necessarily brief, account of the main physics methodology courses in print, which are available at present to teachers and methodologists teaching the method of physics in pedagogical institutes.

These courses certainly do not thoroughly cover all the existing material on the method of physics in the USSR. Apart from these courses there is also extensive material in the form of individual books, articles in methodology journals, and thesis papers on the method of physics. In the following chapter we shall endeavor to give a description and an analysis of this valuable material.

7. Before going on, mention should be made of F. I. Ivanov's papers, which also form a complete system of methodological views, presented in the form of a manual, which for ten years remained as a manuscript. These papers characterize most clearly the innovations which the Soviet school introduced into the teaching of physics.

Ivanov's basic pedagogical idea was that the physical notions which are presented in the form of a systematic plan are grasped by the student much more easily than they would be if studied unsystematically and in isolation. It often happens in teaching practice that the exposition is made up of self-contained stages, without any logical link between them.

Proceeding from this basic idea, together with his vast personal teaching experience and observation of teaching procedures, Ivanov produced in 1939—1941 an outstanding methodology course. He presented this course in a series of lectures, which was reproduced in duplicated copies. The techniques developed by Ivanov at the Research Institute of Secondary Schools and the School Research Institute were applied to experimental teaching at a number of Moscow schools and, more important, at a number of schools in the outlying districts. At the end of the year the teachers met in conferences and summed up their findings. The working procedures proved for the most part quite successful, but it must be said that almost none of the instructors made complete use of the proposed procedures.

The teaching profession lent its full support to the author's basic idea, to the advantage of students, who found their physics studies greatly facilitated. A careful check of the results of teaching by the proposed method showed conclusively that it was possible not only to cover the material of the curriculum but even to exceed it.

The level of physics instruction attained by the students leaving school stands in direct ratio to the quality of the method of physics, which determines, on the whole, the quality of the teacher as well. This may be verified by comparing the Soviet courses (which have their own faults) with Mann's booklet and even with Karl Hahn's German course.

The Soviet courses seem to give a much more complete body of knowledge than the foreign ones. Mann presents the reader with some thought-provoking ideas, from which, however, sufficiently practical results cannot be immediately derived. Karl Hahn thoroughly discusses the specialized aspects of methodology but he does not provide an integral picture of the teaching of physics.

III. THE CURRICULUM IN THE METHOD OF PHYSICS IN PEDAGOGICAL INSTITUTES

The physics methodology courses described above have been written with due consideration of the established and approved curriculum in the method of physics current in the pedagogical institutes.

Since its inception in 1933, this curriculum has undergone several amendments.

The curriculum of the course and of other methodology classes should cover all aspects of teaching, and train as thoroughly as possible the teacher-to-be in a short time, within the framework of university studies.

The studies are based on a course of lectures which present all the aspects of the method of physics which the future pedagogue might require.

This course is rendered more valuable by the fact that it is constantly revised on the basis of practical experience.

In addition to attending the lectures, the pedagogy student does practical work on experimental techniques and takes part in teaching practice. We cannot at this point go into these particular studies, but we should mention the fact that they have to follow the same line as the lecture course. Specifically, the role assigned to experimentation in the teaching of physics in schools must be in full harmony both in the course and in the practical work.

The development of curricula in the method of physics follows a line of continuous improvement without any abrupt changes. Now that the whole constitution of the school has been changed and that polytechnical training has been put into full effect, the time may have come for fundamental changes to be made in the curriculum of the method of physics as well.

The curricula are the results of teamwork and they are more representative of the actual state of the method of physics than the courses. A curriculum which has been approved by the authorities is binding for anyone teaching the method of physics.

The first curriculum which we are going to consider was approved by the State Academic Council on 1 February, 1933.

This curriculum was composed by A. V. Beryshkin, P. I. Popov, and I. I. Sokolov. According to the curriculum the studies in the method of physics are classified under three headings, as mentioned above, viz., theoretical studies (lectures), work in a "methodological" laboratory, consisting mainly of training in experimentation, and teaching practice "in educational institutions".

This combination of three kinds of studies prevailed up to the forties. Only the specific content of the theoretical course and of the laboratory studies was modified to a certain extent. More recently, in the fifties, there has been a tendency to transfer some of the problems formerly pertaining to the lecture course over to the laboratory-seminar studies, i.e., to translate them into practical work. We shall have more to say about this further on.

The general purpose of the studies is defined in an explanatory note to the curriculum of 1933 as "training the students for the teaching profession, on the basis of scientific data." (Emphasis mine.–D.G.)

The curriculum of 1933 is rather abstract in its formulation, but it develops some quite valid methodological ideas.

The expression "teaching profession" correctly indicates the purpose of the studies, but it does not resolve the problem as to what is to be studied in the lecture course. Similarly, the expression "proceeding from scientific data" may refer either to physical facts or to a scientific pedagogical treatment, i.e., it presupposes the existence of a method of physics as a specialized pedagogical discipline.

In the curriculum of 1947 the purposes of methodology are formulated more explicitly, but hardly any better than in the curriculum of 1933.

The curriculum of 1947 states that:

The method of physics establishes (1) the aim of physics education in school; (2) the subject-matter of teaching within the given scientific domain; (3) the set of procedures and methodological means to be employed in order to impart to the students the material in the physics curriculum.

These propositions are quite true, but they are applicable to any methodology and are in no way indicative of the specific aspects characterizing the teaching of physics.

This faulty formulation may be explained by the fact that the authors of the curricula have made no effort to lay the foundation to the method of physics by proceeding from the characteristics of that science and the aims of the Soviet school.

The formulation of the second section of the 1943 curriculum, "Fundamentals of teaching a physics course," misses the point, too. This section deals with physical theories but leaves their methodological implications in the secondary-school physics course unexplained. It would appear that the information supplied in this section should precede the discussion of the aims of physics teaching, as these aims derive from the implications of the physical theories.

The largest number of hours (48) of the lecture course in the 1933 curriculum is allocated to "Physics teaching methods." The term "method" in this formulation appears unwarranted, since it rather involves teaching procedures. The section came to be called subsequently "Aspects of the organization and methods of teaching" but was eventually removed altogether.

The section dealing with the methods of the different topics in the secondary-school physics curriculum has been very superficially formulated.

The curriculum of 1933 develops the main lines of the physics course, but it is not always clear what exactly the lecturer is supposed to explain in such topics as, say, solid-state mechanics, or electricity. The section, which deals with specific topics, was expanded on subsequent curricula and it was stipulated that the lectures should cover all the topics on the curriculum without exception. It is, however, virtually impossible to meet this requirement in practice. It has been mentioned above, in the review of

Znamenskii's physics methodology textbook, that methodology cannot be made clear by any brief exposition of the methods of a specific topic. The same would apply to the lecture course as well.

The next curriculum was issued in 1936 and contained several changes from the 1933 curriculum.

The 1936 curriculum was composed by another team, which included only I. I. Sokolov from the authors of the 1935 curriculum. The new team included two representatives of the K. Liebknecht Pedagogical Institute — E. N. Baradansin and S. I. Ivanov, together with A. V. Pavsha and A. A. Torchinskii; the curriculum was edited by A. N. Zil'berman.

At the same time the Liebknecht Institute set up a specially designed "methodology laboratory," as it was called in the 1933 curriculum. The Institute worked out a complete system of training in experimental techniques, designed to familiarize the pedagogy student with both laboratory and industrial procedures. This system was introduced into the second section of the 1936 curriculum.

The program is much more explicit about the "Aspects and methods of organization of studies." As before, the section dealing with individual topics is not very specific but it contains a significant innovation. According to the instructions in the curriculum, the lecturer must correlate any particular section in the first and second stages of the course. Thus, the curriculum gives "Mechanics for the 5th and 8th grades," "Electricity for the 7th and 10th grades."

It is interesting that in the following program of 1938 principal attention is given to the standpoints from which the lecturer must consider each topic in the curriculum. The 1938 program quotes the following standpoints: the purpose of the topic under consideration; criticism of the methodological procedures not to be adopted in the consideration of a topic in order to avoid possible mistakes; the means of keeping records of the topic; the suitability of the topic for the classroom. The lecturer is thus saddled with painstaking details. This kind of elaborate treatment was already encountered in Goryachkin's methodology. There it was of definite value, even though it made the course excessively bulky. However, it is hardly expedient to have all the details included in a lecture course. These particular points, which are of practical importance in the consideration of any individual topic on the physics course curriculum, have also been mentioned in the explanatory note to the 1956 curriculum, composed by A. V. Peryshkin. In this curriculum, though, the main attention is devoted to the basic aspects of polytechnical education.

Let us now examine one of the recent curricula — that of 1950 — in more detail. The curriculum contained the following sections, denoted by letters: A, a lecture course; B, practical courses in which seminars are held on questions which are difficult to present in lectures, such as the planning of courses, the selection of problems, the perusal of textbooks, and the organization of inspection tours. This project is quite interesting but difficult to realize, as it demands from the pedagogy student a sound knowledge of the teaching procedure. There is no doubt about the practical value of this form of studies, though it actually involves training in the art of teaching.

Under the next heading, C, come laboratory courses; this section specifies two kinds of courses: attendance of the classes of veteran teachers and their analysis, and courses on experimentation in physics. The content of the course remains basically the same as that laid down by the methodologists of the Liebknecht Pedagogical Institute.

The attendance of classes held by other teachers and their analysis is also a training procedure in the art of teaching and is, of course, of tremendous value for the future pedagogue.

Lastly, heading D covers pedagogical practice and presents specific problems of its own. Teaching practice is used to check the pedagogy students' grasp of the methodological-theoretical propositions.

The 1956 curriculum contains a fairly detailed explanatory note, which elaborates at length the specific aspects of headings B, C, and D, and dwells only in brief on the content of the lecture course. This plainly shows the "practical slant" determining the whole organization of the courses on the method of physics by this curriculum.

It was not judged necessary in the 1956 curriculum to develop the method of physics as a theoretical pedagogical discipline. No attempt was made in the curriculum to provide the pedagogy student with some generalized methodological "principles" which would enable him to draw practical conclusions for every concrete case. This bias is clearly seen by going over the topics specified in the course. The section contains sixteen different topics, varying in content and importance. The first heading covers the "Method of teaching physics and its problems. Methodological literature", and then, under the second heading comes the "Educational importance of physics in school."

It would have appeared more natural to follow an inverse order of presentation, i.e., to work out methodological rules and the importance of physics in the Soviet school, proceeding from an ideological and polytechnical standpoint.

It is interesting to note how formalistically the explanatory note to the curriculum defines the problem of teaching physics; thus: "The study of a clearly circumscribed (an unusual expression, often occurring in the curricula of different years. —D.G.) body of systematized knowledge; the acquisition of proficiency in applying the knowledge; acquaintance with dialectical-materialistic views, and the inculcation of Soviet patriotism."

These objectives have applied to the teaching of physics ever since physics became an independent school subject, in 1931-1932. But why should emphasis be placed in the curriculum on a c i r c u m s c r i b e d body of knowledge; why should the problem of teaching physics as a general educational subject be separated from the problems of polytechnical education, which are considered under the following, third heading? Could there be any contradiction between what must be taught in physics, proceeding from a "clearly circumscribed body of systematized knowledge," and that which must be taught, proceeding from the purpose of polytechnical education? One should think not, but then why does the curriculum separate these two aspects?

Under the same list of headings we encounter "on an equal footing" a wide variety of questions, such as teaching procedures and "the material means employed to impart to the students the material on the physics curriculum."*

Included is also a consideration of the equipment of the school physics laboratory, the ways in which classes are to be held, the importance of

---

* Cf. the 1947 curriculum.

demonstrations and laboratory work, visual aids, e.g., motion pictures, textbooks (coming after motion pictures!), the solution of problems, excursions, methods of reviewing, tests, and, finally, studies outside the classroom. There is a total of sixteen such items, for which the lecturer allocates sixty-eight hours over three semesters.

It is thus quite understandable that the pedagogy student often feels dissatisfied after having attended the lecture course offered to him on the method of physics. This course cannot impart the art of teaching, but only provides the "fundamentals" of pedagogical science, which is known as the method of physics.

The effectiveness of the lecture course can certainly be improved, to the extent that it is correctly and systematically constructed. This means providing the pedagogy student with a set of tested principles, by means of which he should be able to solve on his own a variety of problems encountered in the course of teaching.

The compilers of the 1938-1945 curriculum had not discarded the possibility of providing within the lecture course the elements of some kind of theoretical method of physics, though they were perfectly aware of the fact that this field still required elaboration and that much in it was purely empirical as yet.

Now the 1956 curriculum, as we see it, denies on principle the possibility of a logical classification of the material on the method of physics required by the future teacher. This is its main drawback, and it is not made up for by the number of valid ideas presented in the explanatory note. These ideas pertain to the connection between physics and technology. The explanatory note states that "in oder to engage in work in physics one should be acquainted with the basic achievements of technology, and, conversely, creative work in any technological field is impossible without a sound knowledge of physical principles." These quite correct ideas stand in certain opposition to the suggestions made in the curriculum as to how the lecturer must organize the presentation of the methodology course.

The 1956 curriculum presents the advantage that according to it the studies on the method of physics are not restricted only to the attendance of lectures. The curriculum provides for six different ways in which the pedagogy student is taught the methods of teaching physics. These are as follows: the student attends lectures and sits for an examination, he takes part in practical courses where he plans the material that he would eventually have to teach, and he attends the classes of older teachers and benefits from their experience. He learns how to conduct demonstration experiments, to give laboratory assignments, how to work with a saw and file, and to build physical instruments. The practical course in pedagogy involves also teaching for some time in school, and, finally, the student has to write papers on methodological subjects.

It is obvious that if all these various training procedures are suitably conducted and follow one another in the proper order, they will provide the young physics teacher with a good background. This practical trend in the training of pedagogues is particularly important nowadays, when polytechnical training is more and more systematically put into practice in school.

This, however, does not make the drawback of the curriculum noted above, i.e., its theoretical inadequacy, any less important. Any training effected by practical means can be rendered more effective when supported by definite generalizing concepts. A neglect of theory is not consonant with our times, when in many subjects of physics, especially in nuclear physics, considerable practical results have been achieved through the application of most abstruse theories.

IV. SOURCES FOR THE CONSTITUTION OF A SCIENTIFIC METHOD OF TEACHING PHYSICS

1. The five sources of the scientific method of physics

For the constitution of a scientific pedagogic discipline, in our case the method of physics, we may note five sources, inadequately utilized at present. Their synthesis may yield a really new scientific method of physics. These sources are the following:

1. We have mentioned before that the intensive development of the Soviet method of physics is contingent on the development of techniques for the training of future physics teachers at a large number of pedagogical institutes. With the appearance of the first pedagogical institute, it was necessary to collect together the main methodological resources in the big centers (Moscow, Leningrad, Kazan, Kiev) and to give guidance to the local methodology teachers, to train them to some extent in this then little-known field.

Times change, however, and the large number of scientific workers and teachers in the pedagogical institutes now make available the necessary resources for the development of a scientific method of physics.

Thus every lecturer and professor of methodology must carry out a considerable amount of original work, which can provide generalizations and results quite valuable for the scientific method of physics. This remains at present a virtually untapped source.

Unfortunately, in spite of several attempts to arrange a conference on the method of physics for these scientific workers (c.f. the interesting note written by I. I. Sokolov and S. I. Ivanov, on the conference on the method of physics which was to be held by the GUUZ* of the Ministry of Education in January 1956), such a conference has not yet taken place.

The views and achievements of all these workers on the method of physics are almost unknown to us. They are only represented in individual articles they have written and theses they have presented.

2. Another source are, no doubt, the courses on the method of physics written by experienced authors. These courses provide a synthesis of the available experience of others and of the author's personal experience in the teaching of physics in school and of the method of physics in pedagogical institutes.

We have already dealt at sufficient length with the content of these courses and the methodological ideas presented in them.

3. A third source is the literature on the method of physics, which is constituted in a certain measure by monographs, and also the large number of articles in the journal "Fizika v shkole" (Physics in School), as well as articles in other pedagogical journals.

We may include in this source, as an individual category, the teachers' reports in the "Pedagogicheskie chteniya" (Pedagogical Readings) at the

---

* [Glavnoe upravlenie uchebnymi zavedeniyami — (Main Administration of Educational Institutions).]

Academy of Pedagogical Sciences and in the regional institutes for the professional training of teachers.

We shall give further below an individual, brief review of the literature in this specific field.

4. The fourth source, which is becoming increasingly important is the large number of theses on the method of physics presented for the degree of Candidate of the Pedagogical Sciences.

We may place in this classification the drafts of scientific research institutes, which can be described as the treatment of methodological questions. We shall have occasion to return to this particular point further on.

5. Lastly, as the fifth, and possibly most important source, we must consider the experimental work done at the model schools of scientific research institutions and of the departments of pedagogical institutes.

The base school constitutes for a scientific research institution a laboratory, where pedagogical experiments can be carried out both under "natural" and under controlled conditions.

The conditions offered by the base school are best suited to the solution of methodological problems. The experiments must, however, be so conducted as to ensure the reliability of the obtained result.

The importance and limitations of pedagogical experiments will be further discussed at the end of this paper.

2. The method of physics in monographs and journals

We shall try to describe in broad outline the method of physics as it emerges from the journals and the specialized literature, mostly in the form of monographs.

The first characteristic of this literature is that it is directed at the working teacher rather than the pedagogy student. The methodology presented in this literature thus exhibits features which are closely related to real life. The author of a paper or article — mostly a teacher himself — addresses himself to his fellow-teachers, and is more concerned with the concrete conditions of teaching than the methodologist delivering a lecture on the method of physics. Even when the author falls into error, his blunders do not carry that academic flavor characteristic of the errors committed by the methodologist, who is concerned with producing an integral methodology course.

The suggestions given in methodological articles are usually much more detailed and specific; they provide more particulars on the actual teaching procedure, as directly related to the author's personal experience. The result is that often procedures are suggested which may be of value only in the classes of the author but have no objective, general validity. This is a weak point of many articles which never rise above a purely empirical approach in the methodology of any particular aspect. It often happens, therefore, that the suggestions given are quite difficult to follow. However, there are many articles in which an interesting theoretical methodological idea is presented, which another teacher may be able to adapt to his own needs and thus improve his teaching technique.

The method of physics is often charged with subjectivism, i.e., with the lack of "scientifically established" methodological solutions adapted to any set of circumstances, if one may say so, "for all times and all peoples."

By analyzing the articles in journals one frequently notes, unfortunately, that they fail to present a methodological theory. On closer examination, however, it turns out that it is unjustified to expect methodology to provide a solution to every specific aspect of teaching. Methodology cannot think for the teacher. When the teacher is in class, his own viewpoint is the ultimate arbiter.

We started our exposition by mentioning the great need for a methodology deriving from the mass scale on which physics is being taught; it is this mass scale which requires and also makes possible a scientifically established method of physics, but owing to this very fact the directions of methodology assume a sort of "statistical" character.*

It thus appears to us that the future development of methodology, which should be effected mostly through the journals and the specialized literature, ought to concern itself much more with the theoretical aspects of the method of physics than is presently the case.

A basis for methodological formulations may be the deeper analysis of elementary physics, with a view to rendering more precise those aspects which are left aside on a higher level of teaching. The need for such an analysis may be illustrated by the broad debate which took place in 1954 – 1955 in the pages of "Fizika v Shkole" (School Physics) in connection with the articles of Professor N. N. Malov on the obsolescence of many concepts and of D. D. Galanin on the concept of mass.

These theoretical methodological researches may be directed towards a closer relationship with the psychology and physiology of children and adolescents. Such a trend is at present almost non-existent in the physics methodology courses.

Quite a large number of the articles in the methodological journals is usually devoted to experimentation. It is often in setting up a new experiment or an original version of an old one that the teacher can most naturally concern himself with the methodological aspects of physics.

No systematic attempt has yet been made to interpret the bulk of material in the methodological journals, though it would prove of considerable interest. Of course, every worker in the method of physics and almost every teacher reads these journals and their contents have a certain influence on their views and their practical work. P.A. Znamenskii's methodology, more than any other, exhibits the influence of the monographs and journals.

Unfortunately, the specialized monographs on the method of physics appearing in our country are relatively few, and the topics which are treated are picked more or less at random.

These are usually either detailed methodological expositions of a particular topic (for instance, P.A. Znamenskii's book "Voprosy volnovoi teorii sveta v kurse fiziki srednei shkoly" (Aspects of the Undular Theory of Light in the Secondary-School Physics Course), or K.N. Elizarov's "Elektromagnitnye kolebaniya i volny v kurse fiziki srednei shkoly" (Electromagnetic Oscillations and Waves in the Secondary-School Physics Course), or else books dealing with some particular teaching technique,

---

* The term statistical must not be interpreted merely as an "averaging," but rather in the more profound sense of principles for drawing inferences from a large number of facts, analogous to the principles of "physical statistics."

e.g., laboratory assignments (V. N. Bakushinskii), of the graphical solutions of problems (L. I. Reznikov). These monographs do, of course, develop individual elements of methodological science, but they are unfortunately still far from forming a self-contained, scientifically consistent system of physical methodology.

We would like to mention specifically the books on physical experiments written by A. A. Pokrovskii in conjuction with some other scientific workers. These books present the principles of a new methodology of physical experimentation. The distinctive feature of this methodology is the fact that the various experiments are planned with a view of the didactic purpose for which they are designed. The authors also indicate the minimum necessary equipment, which should prove inexpensive to acquire. These valuable books are mostly concerned with practical issues, and the theoretical aspect of the method of experimentation is presented in a very condensed form, though it has been developed at full length by A. A. Pokrovskii in a big, unfortunately as yet unpublished, book on physical experiments.

Accordingly, the journals and monographs do not at present give a well-rounded picture of the Soviet method of physics, but are mainly devoted to the development of individual details.

The development of these details is often quite apt and valuable. The correlation of the material in the monographs and especially in the journals and its critical evaluation constitute one of the topical problems of the future method of physics.

3. The method of physics in thesis papers

The number of these presented on the method of physics for the degree of Candidate of Pedagogical Sciences attains at present some hundred and fifty.

The theses presented on methodology go through a long process of careful checking before being approved. Their importance for the progress of methodology can therefore be hardly over-emphasized — it is very high indeed.

Of course, the same drawback which was noted in the journal literature, i.e., concentration on the topic under discussion, almost to the exclusion of the general situation in the teaching of physics and of the general problems of methodology, is quite manifest in the theses. For instance, the theses dealing with the individual topics on the curriculum always attribute to the latter more scope than they could possibly have if due allowance is made for the general level of the physics course in secondary school. As a result of the desire of every author to say all he can on his topic, many pages of the theses turn out to be just a compilation of facts, on a higher or lower level. The compilation of facts may give substance to most theses, but it does not particularly help in the scientific formulation of a physical methodology. Any scientific synthesis constitutes in a way a chain reaction, in which the results of one work produce fresh problems to be resolved. As is known, for a chain reaction to be formed the "chains" must not be broken and the "multiplication factor", defining the rate of the reaction, must always exceed unity. Now in the thesis papers on methodology which constitute, in a manner of speaking, fragments of the scientific method of physics in its pure form (i.e., almost free of pedagogical problems directly pertaining to teaching), it is very rarely seen that the author tries to draw on the work of other methodologists. On the contrary, the author usually prefers to start his synthesis from the very beginning, working in a vacuum as it were. Such an approach to research is quite untenable at the present time, and the author of a thesis is in a position where he has either to challenge the findings of his predecessors or else to accept them and proceed from the ground already covered. However, such debates, whether pro or con, are usually held on specific issues only: the general, fundamental problems of the method of physics as related to secondary education still remain largely unclear. In point of fact, a large amount of material has been already accumulated in the thesis papers, which only requires systematization and correlation in order to make it conducive to the creation of an objective, generally valid, scientifically derived method of physics.

The correlation of the thesis papers is made difficult by the fact that very few of them are published. If the thesis papers were duly published they could be made the object of a general discussion, which would obviously lead to a clearer formulation of the basic principles of the Soviet method of physics.

The thesis paper may be divided into three categories, specified below, beginning with the one containing the largest number of papers.

We will refer to the first category the papers dealing with the methodology of the different topics on the secondary-school physics curriculum.

A particular kind of standard procedure has come to be adopted for these theses: a historical review and the contemporary state of the subject in science; proposed methods of exposition; some findings of practical application, and final conclusions.

In the first theses of this kind (e.g., V. F. Yus'kovich—"Brounovskoe dvizhenie" (Brownian Motion), L. I. Reznikov — "Volnovaya optika" (Undular Optics), E. E. Evenchik —"Poverkhnostnye yavleniya" (Surface Phenomena)) the applicants aimed to achieve a two-fold practical purpose: not only to provide methods for the study of a given topic in secondary school, but also to collect and process the physical facts which would be required by the teacher in order to present the subject well. For this reason the first part of the thesis, which fulfilled an independent and important role, constituted to all intents the main point. In the better samples of this kind of thesis (cf. Yu. I. Sokolov's thesis on work and energy) the applicants have made an effort to regulate the "rear-guard" of physical science, i.e., to formulate some new, more precise definitions of physical concepts which would present definite methodological advantages. Unfortunately, now that there is a plethora of theses and that the curriculum subjects seem to have been exhausted, theses of this kind have lost their merit, their content has become formalized, and the scientific-methodological quality has dropped.

The last extensive work carried out on the physics curriculum in 1951 – 1954 by the Commission of Curricula of the Ministry of Education showed the low practical effectiveness of the large majority of these papers, to which the members of the commission paid almost no attention. This was due mainly to the fact that most of these specialized theses remained unknown or that their findings were forgotten.

The critical review and correlation of these numerous theses could well make the subject of a special work, which would fill a definite need.

The second category, which is considerably less extensive, covers papers of a pedagogical-historical content. These are concerned either with individual eminent methodologists (for instance, the thesis on A. V. Tsinger) or with the different periods in the development of the method of physics.

The value of these papers can hardly be denied, of course provided they do not dwell on historical trivia, but also reflect the development of methodological ideas and the relationship of this development with social classes. Many valuable ideas of the past are often completely forgotten, and later re-discovered, sometimes after considerable labor. Also in the case of these papers it is very important that they should be published.

Lastly, in the third category we classify the papers devoted to methods of teaching and other aspects of varying content. At one time, for instance, many theses dealt with the solution of problems, another group is constituted by the theses on assignments outside the classroom (K. A. Shakhova), the application of motion pictures (S. I. Ivanov), etc. To the present category are also referred the theses entirely devoted to new experiments (A. I. Glazyrin, A. G. Dubov, etc.). The value of this group of papers for the general method of physics could have been much higher if there had been more accordance in the authors' consideration of various aspects, and if they had proceeded from the findings of one author and elaborated on them. This is the only true scientific work, similar, as we said, to a chain reaction, which proves really fruitful. Unfortunately, there is almost never such continuity, even when a fairly large number of theses deal with the same subject, for instance the solution of problems in physics.

Each author of a paper on this subject tackles it to a large extent subjectively; for this reason, for instance, this particular subject of problem solution in physics cannot be considered thoroughly investigated, in spite of the number of papers devoted to it. With the exception of some unquestionable elementary propositions, the authors of major methodology courses take almost no account of the results of the many thesis papers when they proceed to present their methods of problem solution.

A very large proportion of the thesis papers written before 1954 does not take into account the tremendous advances made in the method of physics through the implementation of polytechnical training; it is only in the papers of the last year [1958] that distinctly new viewpoints are tentatively beginning to emerge.

Thus, notwithstanding the shortcomings of the thesis papers, some of which we mentioned, this material constitutes a rich fund of living methodological thought and the correlation of extensive experience in the teaching of physics in our schools. Most of the authors of the papers are themselves physics teachers and have had the occasion of examining the teaching procedure while working on their thesis. The theses indubitably offer an interesting synthesis of living school experience, and it is unforgivable that this valuable material remains completely foreign to a large circle of teachers and methodologists.

V. THE PHYSICS TEXTBOOKS AND MANUALS AND THE SCHOOL PHYSICS LABORATORY

1. The successful teaching of physics is determined by three main factors, and it is difficult to say which is the most important one, as their effects on the student are very closely related. These factors are: the teacher, the equipment, and the textbook.

These three factors are a subject of study for the method of physics.

Among the many imaginative teachers, who managed to impart to their students sound polytechnical knowledge (out of the several thousand we will mention only two — I. I. Babushkin in Chapaevsk and N. N. Shishkin in Baku), there are in our schools also teachers who teach physics in a most formal manner. Under the existing conditions they are unable to cope with the difficulties they encounter, and the students at these schools are left with inadequate knowledge. It is true that physics is not one of the subjects in which many student failures are noted, but this is, regrettably enough, due to the fact that less importance is attached to the knowledge of physics than of spelling, say, rather than the fact that the graduates of the 7th or 10th grade possess particularly sound and extensive physical knowledge.

All the teachers, whether poor or good, are badly in need of a physics textbook which would constitute a useful aid in any of the difficulties that may crop up in the course of teaching. The members of the teaching profession were almost unanimous, when their qualified opinion was solicited, in stating that the standard textbooks failed to meet this requirement.

In the methodology courses practically no appraisal is given of the textbook, and there is just the barest minimum of references to the standard textbook. In E. N. Goryachkin's methodology, for instance, the textbooks of the 6th and 7th grades are almost not mentioned at all.

This disregard of the textbook in the methodologies stems from the fact that the negative aspects of the textbook are quite plain to every methodologist, whereas its positive aspects are difficult to define.

The methodologies we have considered and the monographic literature are, obviously, of a much higher quality than the textbook. Unfortunately, though, the students study physics from a textbook, and not from a methodology, and certainly not from individual good books.

This situation is quite untenable, and it is impossible to think of any appreciable improvement in the teaching of physics without a r a d i c a l i m p r o v e m e n t of the physics textbook for all grades, from the 6th to the 10th. We cannot dwell here on this sore point in full, but we may unreservedly state that the quality of the current textbook is b e l o w the standard set for physics in a school where polytechnical training is to be achieved.

2. Physics cannot be studied only from a book, no matter how good the book may be. A problem of prime importance thus arises — that of the equipment of the school physics laboratory, throughout the Soviet Union. The equipment of the physics laboratories involves not only considerable financial expenditures; it is also necessary to establish a standard sample for each instrument and to produce the instrument in large quantities. For the time being, despite the extensive and fruitful work performed by A. A. Pokrovskii and his colleagues at the Academy of Pedagogical Sciences, we

do not have a definitive minimum set of standard physical instruments to be used for pedagogical purposes in polytechnical training. The specification of such a standard set of instruments as well as of the other equipment for the school physics laboratory is one of the main tasks of the method of physics.

But specifications and standard samples are not enough. It is also necessary to modify the industry of school equipment (possibly by making use of the valuable experience of our colleagues in the German Democratic Republic and Czechoslovakia), to enable it economically and efficiently to supply instruments for all our schools.*

3. Let us now turn back to the physics textbook, in view of the importance of this problem. On the basis of the foregoing, let us formulate the requirements which have to be met by a good physics textbook:

a) The textbook must present the principles of modern Soviet physics and provide the students with a well-integrated physical outlook, based on the ideology of dialectical materialism.

In this way a balance will be struck in the textbook between the purely experimental and the theoretical presentation of physics and there will be an adequate discussion of the technological applications of physics. The student will be also able to pick up from the exposition the generalized propositions which should enable him to apply his knowledge in practice. We particularly stress that the physical knowledge provided by the textbook must not be split up into individual propositions which are loosely connected with each other, even though this is precisely the way in which physics is presented in many of our standard textbooks. The textbook must provide a "system of viewpoints" which are related to modern physics, a "system of concepts" or a "world picture" which gradually unfolds with each section and is integrated into a physical outlook by the relationship linking these sections together. For instance, the ideas on the three states of matter must be most fully presented in the study of molecular kinetics and be supplemented in the study of atomic structure.

b) The textbook must give (especially in the first years of physics study) a fairly detailed and descriptive account of experiments, and the explanations must be thorough. The experiments may be performed either by means of common household items, or by means of the standard minimum equipment of the school physics laboratory, which was mentioned above. In the elementary physics course the textbook must make extensive use of photographs of the instruments and setups (preferably in color), in conjunction with line drawings and diagrams. The experiments would be best represented by a series of pictures showing them in the process of being performed, i.e., five or six drawings or photographs depicting various successive stages. Further on in the exposition the description of experiments may be made shorter.

c) The general trend of the presentation must be largely inductive, i.e., proceeding from the description of the phenomena to their investigation and explanation. After the phenomena have been investigated, the requisite

---

* Let us add at this point that the maintenance and operation of the equipment in the schools is quite bad, so that instruments are put out of commission within a very short time. This unfortunate situation is due to the poor training in experimental techniques at the pedagogical institutes, who do not possess the appropriate equipment either.

physical quantities and the units in which they are expressed are given, and the relationship involved is cast into a mathematical form. In relatively rare cases it is possible, in the senior grades, to subject a formula to mathematical treatment and derive corollaries from it, and then check some of the corollaries by a new experiment.

It is inadmissible to "mathematize physics", as it is a mistake to think that physical laws can be "discovered" by mathematical means.

The exposition may in some cases follow deductive lines, where special results are derived from general propositions. For instance, after establishing by experiment the law of refraction, it proves possible to construct by deduction the path of a ray through a prism, a parallel plate, and a lens, and only later check these conclusions by experiment. It is in no way admissible, however, to adopt a manner of presentation which accustoms the student to a dogmatic way of thinking. Every statement or proposition with which the student is presented must be supported by the proper foundation or proof.

d) The presentation in the textbook should be divided into individual sections, each comprising a self-contained part of the exposition. The sections are joined together into chapters and the chapters into parts, thus showing clearly the structure of the system of concepts, whose importance was noted above. The language in the textbook must be simple, clear, and grammatically correct. Descriptive comparisons are in order, to the extent that they make things clearer. Similes and metaphors should be judiciously chosen and used in moderation.

The style of exposition will depend to a large extent on the subject treated, whether an experiment, the derivation of a formula, or the description of a theory. In this connection, it would be desirable to have at the end of the various chapters and parts a section in the nature of an essay or a popular-science article, which either generalizes the material previously presented or gives a clear account of physical theory or of technological applications.

In the textbook for the first stage of the course these generalizing articles can be written in an informal style.

This suggestion as to the arrangement of the book is made with specific purpose in mind.

The point is that, considering the precise and condensed style of the textbook, it is difficult to present physical theories and various technological applications in a straightforward manner. The articles to be introduced in the textbook in the form of essays should make easier the exposition of such aspects and increase the polytechnical value of the textbook.

There have been some attempts in recent years to abridge the textbooks. It was argued that this abridgment would reduce the amount of homework imposed on the student. If, however, the material in the textbook is separated into the material for study and that for straight reading, this argument loses its force. On the other hand, abridging the textbook from considerations of economy is on no account admissible if it reduces its quality.

e) One of the principal purposes of the textbook must be not only to provide the student with the material he would need for his examination but also to get him used to applying the knowledge acquired.

The whole manner of presentation should be such as to contribute towards this end. The student should be able to find in the book examples of applications in the description of technical physics. In addition, the textbook should contain a number of problems (sometimes with solutions) showing how to apply the information presented. It should be added that a great many of the current books of problems in physics serve this purpose rather badly.

f) The textbook must be the first and basic book from which the student studies physics, though for the purposes of polytechnical training and for a better grasp of physical facts it would be quite useful for the students to have a library which would supplement and enlarge upon the contents of the textbook, at the same time being in full accordance with it.

g) In the polytechnical school the subjects of study must be closely correlated. It is inadmissible that any subject should be isolated from the others. Thus the physics textbook must be in full accord with the textbooks in chemistry, mathematics, geography, biology, and even history. The scientific materialistic outlook imparted to the student in the school should be achieved by the concerted action of all the subjects.

h) The authors of physics textbooks should keep in mind the fact that there are in school also workshops and technical laboratories, so that both the text and the exercises in the book should contain examples related to these activities.

The other school books, i.e., the book of problems in physics (edited by P. A. Znamenskii) and the specification of laboratory assignments (by P. A. Znamenskii and A. A. Pokrovskii) are in a better condition than the textbook and their improvement is not a pressing task. We can therefore pass over the methodological requirements which these books have to meet. One drawback should be noted, however: there is a certain lack of correlation between the various study books in physics that are put out.

We mentioned the fact that the textbook must not be the only book from which the student derives his knowledge of physics. It is necessary to have reading books for the student which are correlated with the curriculum and with the textbook. The need for this type of books is becoming particularly important at present, when the physics course must be the basis on which the student becomes acquainted with concrete technological applications. We spoke before about separate chapters in the textbook to be devoted to essays explaining technological applications of physics. Such chapters are certainly not enough. The student, while going over the physics course "by the book," must also read all kinds of popular-science literature. This literature should compose the students physics library. One of the tasks of teaching physics is to teach the student how to read books on physics and technology.

The method of physics thus has the problem not only of determining the quality of the textbook (we tried to indicate these briefly above), but also to work out the methodological requirements to be met by the content and style of exposition of the student library.

It must be admitted that very little has been done in the method of physics in this direction as yet.

It may be suggested to issue for the students of the senior classes a special monthly journal, dealing with polytechnical subjects, such as mathematics, physics, chemistry, geography, and biology.

This journal would differ from the existing technical journals ("Tekhnika molodezhi" (Technology for the Young — Popular Science) and "Znanie — sila" (Knowledge is Strength) in that it would be closely associated with the school curriculum and with extracurricular studies within the framework of the school.

VI. THE IMMEDIATE PROBLEMS OF THE METHOD OF PHYSICS RELATING TO POLYTECHNICAL TRAINING

We have tried to present the history, contemporary stage, and ways of improvement of the method of physics in the Soviet Union.

The modern method of physics, in its capacity as a scientific-pedagogical discipline, has not yet satisfactorily treated certain problems. The importance of the latter has come to the fore in the process of realizing the polytechnical training program and the tasks set to the school in general, and in the teaching of physics in particular, by the Law on the Consolidation of the Link Between the School and Life and on the Further Development of the System of Public Education in the USSR.

Let us dwell on some of these problems, which appear to us the most important.

1. The applicability and retention of the knowledge of physics acquired in secondary school

The problem of the applicability of the physical knowledge given in school was raised at the very inception of the method of physics, before the October Revolution, by the noted methodologist V.V. Lermantov from St. Petersburg.

Any knowledge in physics is valuable only when the person who has it can apply it in his subsequent activities.

The "applicability" of physical knowledge has to be defined in terms of the acquisition of a physical outlook, the development of observational powers, and the ability of pondering physical phenomena proceeding from the physical knowledge acquired in school. We should not forget also some laboratory proficiency and the ability of carrying out some physical measurements.

"Redundant" or "sterile" knowledge is the purely "school knowledge" which is necessary only in school, merely for answering in class or during an examination, which is by its very nature incapable of being incorporated into the independent thinking habits of the student.

First and foremost "by definition", so to speak, for any knowledge to be applicable at all it must be of a durable kind. Many teachers maintain that even in the case of good students the knowledge on the basic properties of electrical current acquired in the 7th grade is completely gone by the 10th grade. If the knowledge of electricity acquired in the 10th grade will be as transient as in the 7th grade, how is the student expected to apply it after matriculation?

Now how can reliable knowldege be achieved? From general psychological considerations it is clear that reliability is achieved by excercise and application. At the present time, the application of knowledge in the teaching of physics is restricted almost exclusively to the solution of problems in physics.

The application of the knowledge in physics in the solution of the common type of problems is, however, very limited and, in a manner of speaking, one-sided, because it relies only on words and concepts. The problems to be solved are quite often of a stereotyped kind, leaving no room for imagination. Problems involving some kind of "twist," such as are given, for instance, in contests, prove in most cases impossible for the student to solve. The more true to life the problem is, the more difficulty the student has in solving it. Such is the present state of knowledge in physics of the students in a large number of schools.

At its present stage of development the method of physics must set itself as a basic objective to find ways of applying the knowledge imparted in school on a much broader basis.

The applicability of the knowledge of physics, on reaching a certain degree of abstraction, is also contingent on its being a conscious knowledge. Fresh information is then assimilated on the basis of the previously acquired knowledge.

It often happens that the physics course is broken down into individual topics which are not connected with each other. If the topics do not follow from one another, obviously a well-defined "system of knowledge" cannot be formed. A "physical picture of the world" may be produced only when the knowledge from one field of physics is applied to the understanding of another one.

The students can and must be especially taught to apply their knowledge, as if it were just another "topic" in the physics course. A radical revision is due of the requirements to be imposed on the student sitting for an examination or being tested in any other way for his knowledge in physics.

If we devote some thought to these requirements and analyze the specific features of the pedagogical procedure that can satisfy them, we come to the conclusion that almost all the stages in the teaching of physics must be substantially modified.

The principal modification in the pedagogical procedure will thus involve making the student consciously absorb the physical experiment with which he is presented. At the present time we quite often underestimate in the course of teaching the value of our first signaling system, of perceiving physical phenomena "in the making" (V. I. Lenin) and we make very bad use of the exceptional possibilities of the second signaling system that is human consciousness. These possibilities quite often produce in us the "illusion" that the students have "grasped" the physical phenomena being studied. It often happens that the students learn how to "discuss" the phenomena, i.e., to find the necessary words by imitating those of the teacher or in the textbook. What they are saying is in fact quite remote from their independent reasoning. Furthermore, special care should be taken of the formation of abstract concepts from the perception of processes "in the making" and the ideas they produce, especially in the elementary physics course (this is actually the point of the first stage in the teaching of physics).

Mention should be made of the great value that the laboratory work conducted at the end of a half-year or year has in teaching how to apply the knowledge in physics. In carrying out experimental assignments in the various topics of the course, the student has to apply his previously acquired knowledge and compare its various aspects in a completely different way than he had to do it in a verbal discussion. The laboratory assignments have to be designed so as to meet these requirements.

It is at present considered that the preparation for the final examination gives the finishing touch to the learning of the course. Unfortunately, in the great majority of cases the system of examination papers and the current practice of oral examinations require from the student only an effort of memory but not of understanding. This is a bad but still prevalent scholastic tradition, which must be abolished in instituting the program of polytechnical training. The oral interrogation, which constitutes the basis of the examination, can be made to cover a variety of topics, and it may be easily designed to elicit not formal but rather sound, "applicable" knowledge. This is normally not done, however, to avoid displaying the gaps in the knowledge of even the best students.

Training in the "application of knowledge" is also done in excursions and the reports of students and their discussions.

The method of physics must work out a set of measures to be adopted in teaching which would ensure a sound and conscious knowledge, which would thus have practical value.

2. The physics course in its association with the technical sciences

The claim has never been made in the Soviet method of physics that the physics course should be c o m p l e t e l y isolated from the problems of technology and that this course should be given the form of a perfectly abstract scientific discipline.

During the time of "unit" teaching* and at the time of technical application of physics more attention used to be given to the problems of technology than in the thirties, after physics had been established as an individual school subject.

At that time (the thirties) there were indications that methodologists were trying to divorce the physics course from the problems of technology.

The teaching of physics as part of a "unit" subject naturally led to the study of technological problems. It was these problems, and the fact that they were stated in terms of technology and economics, which determined to a large extent the content of the physics course at the time, so that physics failed to display the problems properly characteristic to it as the "model" science.

Obviously, practising teaching in this way did not lead to good results and could not be kept up for long. The directives issued on the school by the Central Committee in 1931—1932 restored the rights of physics as such, i.e., as the science of the most elementary and yet most widespread phenomena, which underlie all phenomena studied in the other sciences.

When physics was reinstated as an individual subject its method had to give up all teaching procedures in which physical knowledge was built around a technical construct — (let us recall the famous "well" of F.N. Krasikov, the physics of the samovar, the attempt to promote electrical engineering based on the repair of burnt-out fuses, among other).

---

* [See footnote on p.7.]

Methodology had to produce at short order effective teaching methods for a physics course which would be suited to the courses given in secondary and technical schools.

The physics courses in the methodological reference books before the Revolution (e.g., N. V. Kashin, A. S. Baranov) did not contain any technological topics. There were very few of those also in such pioneering works, as, for instance, A. V. Tsinger's textbook "Nachal'naya fizika" (Elementary Physics) or Indrikson's and Grigor'ev's "Kurs fiziki" (A Physics Course) books.

At a certain point* the problem of the technical information to be included in the physics course was subjected to a broad and earnest discussion. For instance, V. Egorshin wrote: "It must be borne in mind that every question (emphasis mine. — D. G.) of a physical nature which happens (! — D. G.) to be relevant to industry should be made to bear on the problems of economics, of the five-year plan, the general electrification plan, technological reconstruction, etc."

"The result of this onslaught", writes I. I. Sokolov in his methodology (1934), "was the appearance, in 1932, of books whose volume exceeded all reasonable limits, at least some of which, as we mentioned before, included more than 50% technical material."

It is quite clear from the above words that I. I. Sokolov is rather hostile to the introduction of technical material into the physics course; he is not in sympathy with the procedure and does not perceive in what ways technology may be introduced in physics. Sokolov's physics methodology does not set itself the task of finding methodological ways of introducing technology into the physics course; the same may be said of the methodologies of the Leningrad authors.

At the present time, however, under the program of polytechnical education, this problem becomes more acute than ever and remains quite difficult as yet.

V. Egorshin specifies that "mentioning things in passing" will not do as the students will not understand, while if anything is explained in specific detail the length of both the textbook and the explanations of the teacher will grow out of all proportion.

This problem can hardly be solved, as I. I. Sokolov suggests (1934),** by conducting a "scientific investigation of the rate at which the textbook is learned." The point is not the "rate at which the textbook is learned," but the fact that the technical account must be designed for other purposes than merely the presentation of scientific views. The technical account in the physics course should provide the reader seeking a polytechnical education with a polytechnical (!) generalization, i.e., with an idea of how a given technological construct is realized in actual practice. Now in the accounts given in the specialized technical literature use is made of special terms, which constitutes part of the systematic exposition; the specialized account is designed for the technician or engineer who uses it to acquaint himself with the construction or the operation of the described construct. For this reason such an account necessarily includes specific

---

* Cf. Egorshin's article "Za marksistsko - leninskoe estestvoznanie" (Natural Science on Marxist-Leninist Precepts), No. 2, pp. 47-49. 1931; quoted from I. I. Sokolov's "Metodika fiziki" (Methodology of Physics), p. 181, 1st edition. 1934.

** Sokolov, I.I. Metodika fiziki (Methodology of Physics), p. 183. 1934.

and precise details, dimensions, specifications of materials, computations of strength, economic aspects, technological implements, etc.

The account designed for polytechnical education should provide a "basic" discussion, without being burdened with details, but neither should it omit giving a technical description. This is to say that a technical account cannot, by its very nature, be unduly generalized; it must of necessity present concrete features, or else it becomes pointless. For instance, any "overall" account of an excavator or an automobile would be of no particular polytechnical value.

In giving an account of an excavator or an automobile one should always keep in view the tasks for which these machines are intended and the various types in which they are produced, and indicate the technical specifications of each type in accordance with the operational requirements it has to meet. Thus, in giving an account of automobiles mention should be made of the difference between passenger cars and trucks, and possibly of the characteristics of the most commonly encountered types of cars.

In a polytechnical account, no matter how short, one cannot help broaching some economic and even technological questions. This gives the account some quite characteristic traits.

On the other hand, any irrelevant, purely technical details (for instance, the makes of machines, dimensions that are subject to variations, etc.) should be omitted from a polytechnical account.

That is the difficulty of polytechnical writing in science popularization.

In the present paper our object is to state the problem, to indicate the difficulty with which the method of physics is confronted, and to try and give some methods of presentation.

We cannot solve this problem here completely.

This problem of bringing the "frontiers" of technology within the reach of the rising generation is actually a recent development, which culminates in the polytechnical training program.

To solve this problem we do not have to start from scratch, of course, as quite a lot has been already done both in general and professional education. It would be wrong, however, not to contemplate the specific features of the problem as it presents itself within the framework of polytechnical education.

Vocational training as well as some other teaching procedures, which fit within the polytechnical education program, exist in schools abroad, and we can possibly borrow some of their methods. We should have, however, to adapt them to our specific pedagogical needs.

The physics teachers both in past years and at present acquit themselves quite well of this difficult task, but they solve the problem "by a rule of thumb," in the manner of an artisan. Now the technology of artisanship is nothing like the technology of modern factory production, and in developing a scientific method of physics we must, of course, provide not a "craftsman's" procedure but a scientifically sound and pedagogically tested solution.

The solution of this problem must be worked out on a high t e c h n i c a l level, so that it should not turn out that the polytechnical knowledge of the students, when they step out into life, is already obsolete and removed from advanced technology. Therein resides a methodological problem of very high importance and of extreme difficulty.

Another task of the method of physics involves making quite clear the point that it is impossible to acquire the physical knowledge necessary for a polytechnical education by studying physics "on appropriately selected technological constructs".

No matter how well these constructs may be chosen (be it an automobile, a bicycle, a tractor, a radio set, a sewing machine, etc.), the physical concepts and laws will be reflected in them only partially, in some specific aspects and not in their general validity. To quote V. I. Lenin, every true and valid generalization (a general concept or law links together several concepts) enables us to understand nature more profoundly and truly than the individual phenomena which were subjected to generalization.

Now if one proceeds on an unsystematic study of physics, then in order to have at the end of the course a knowledge of physics one would have to draw generalizations from the variety of facts, i.e., to study physics all over again.

In particular instances it is of course possible to compromise from pedagogical considerations, but in principle the physics course must be systematically constructed in all its stages and follow the logic proper to science.

"Technological constructs" can and must be studied in the physics classes with a view to developing the students' ability to apply their knowledge.

3. The first stage of the physics course

The law passed by the Supreme Soviet of the USSR on the "Consolidation of the link between the school and life and the further development of the public educational system in the USSR" established two stages in secondary education.

The first of these stages is the eight-year school, in which the physics course should be given its proper importance. The basic content and the methods of study of this course must conform to the requirements of the developing mind of the student.

Thinking in the abstract, it would be more "convenient" to study physics in a single grade, but this should probably prove hard on the student's brain. The student's brain is not a storage compartment but a living, d e v e l o p i n g organ, and the study of physics in the 6th to 8th grades is first of all necessary to enable the student to d e v e l o p. Without this gradual development, associated with the requirements and interests of the student, physics cannot be learned in a "natural" manner without constraint or cramming on the part of the student, prompted as he is to absorb physics by his whole environment.

We are still far from realizing the extent to which "verbalism" is prevalent in our school, the extent to which the material provided for by the curriculum is removed from the needs of a boy surrounded by a reality which is increasingly pervaded by technology. We have already mentioned the fact that the importance of the development of the first signaling system is underestimated. Its value has to be taken into account in the consideration of the elementary physics course.

Up till now almost all the work of the elementary school was reduced to the teaching of reading, writing and arithmetic.

Nowadays, an important part in the development of a youngster must be played by well-designed manual work and workshop assignments.

The student not only goes to school, sits in class, and answers the teacher's questions, but while he is studying at school his mind develops, and he leads a life of his own. It is precisely at the age when a student is in the 6th to 8th grades that the highest interest is displayed towards making things. Therefore, physics must be studied in these grades primarily by the laboratory method, obviously without carrying this method to methodologically absurd lengths. We already discussed this in connection with the methodological system of E. N. Goryachkin.

Modern methodology has the task of developing the already known methodological propositions within the specific framework of the school working under the polytechnical training program, and to find the most effective techniques for facilitating the teaching of physics.

With respect to elementary physics, the following methodological propositions should also be made clear:

1. Knowledge must be acquired on the basis of direct experimentation, and the student should, as far as possible, make himself the equipment necessary for the experiments.

2. However simplified the elementary physics course may be, it cannot be only "qualitative". The experiments must be accompanied by some basic calculations, even with some algebra, and by graphs. This will achieve more valuable knowledge, which the students will be able to apply in real life.

3. Its elementary level notwithstanding, the course should provide some system of the generalizing concepts of modern physics. It must include such concepts as work, energy, charge, the electrical field, and some notions of fundamental physical theories (e.g., the molecular theory, electron theory).

4. The body of knowledge provided by the course of the eight-year school should cover the main branches of physics (in particular, it should contain information on light and sound).

Only in this way can the course furnish the knowledge of physics required in the other subjects.

In our view it was a mistake to have eliminated from the curriculum the subjects of light and sound, as was done in recent years.

Given an adequately equipped school physics laboratory and the possibility of performing a considerable number of laboratory assignments, the physics course should be allotted more hours on the school program for the 6th to 8th grades.

Some thought should be given to the fact that the first stage of the physics course should not be lowered but rather raised, so as to raise the level of the whole physics course.

Can it be considered that all the methodological possibilities have been employed to make the teaching of elementary physics under the present curriculum as effective as possible, or are there still (not inconsiderable) reserves?

It seems to us that there are such reserves indeed, since no one has ever yet tried to teach physics in the 6th to 8th grades by making full use of the teaching techniques recommended in the method of physics, by assigning a sufficient amount of laboratory work, by employing a good, clear textbook, and by utilizing the popular-science literature.

Only such a comprehensive way of teaching can settle the question as to the limits to be set for the elementary physics course.

The students who have been given manual work in the elementary classes and have worked in workshops in the 5th grade are more developed by the 6th grade than the students of the modern school and are in a better position to assimilate physics.

All the members of the young generation have to go through the physics course of the eight-year school before going into productive work.

The productivity of the newcomers will be determined in a certain measure by the extent of the knowledge they have acquired and by its intrinsic quality.

This is how we can formulate the task of the future methodology of the elementary physics course.

4. The second stage of secondary education in physics

A characteristic feature of the second stage of secondary education (the 9th to 11th grades) is the variety of ways in which the school studies are organized.

There are some schools in which the practical training of the students is largely under the control of the school. In these schools the method of physics outlined above is perfectly applicable. The same may be said of the technical colleges.

However, there are schools where the studies are conducted in parallel with the professional occupation of the students, i.e., evening schools and correspondence courses. In such cases the teaching of physics must be organized along different lines from the existing methodological traditions. The method of physics has almost not concerned itself at all with the problem of training by correspondence, although the need definitely existed.

The most important problems to be considered with respect to the independent study of physics are increasing the effectiveness of the textbook and working out new ways of using experiments.

In producing new books for the independent study of physics attention should be given to the methods of exposition employed in the "Rabochie knigi" (Workman's Books) (A. V. Tsinger, E. N. Goryachkin, and others). As for experimentation, provision should be made for an increased amount of laboratory work and for special arrays of physical equipment enabling the student efficiently to perform a large number of experiments.

Many years of experience of working on the physics curriculum of the senior classes and in developing the individual topics in the methodology courses and in specialized theses does not give any indication that the course is excessive in scope or that it delves too deeply into the fundamental aspects of modern physics on the plane of secondary polytechnical education.

On the contrary, we could easily mention more than a score of new subjects which have long been "ripe" for the secondary-school course and are indispensable for the formation of the young generation.

In recent years work on the curricula has displayed a strange contradiction: it has striven to produce a physics course of a higher general educational and polytechnical value, and at the same time to abridge the curriculum.

It appears to us impossible to reconcile these contradictory aims. The polytechnization of the school requires an overall improvement of the physics course, so that it should be possible to proceed from the study of scientific principles to technology, and not the reverse. The level of the physics course must be considerably raised if the school program is made to cover the study of the fundamentals of production and industrial training.

The history of the Soviet method of physics proves, as we have tried to show, the necessity of thoroughly systematizing the physics course. The course ought not to overlook the problems of technology and the applications of physics in industry, just as physics itself deals with technical aspects. After all, physics has engendered many sciences that are purely technical in conception and are the basis for the training of engineers, whose business is the technology of production and the erection of impressive structures. Principles of physics normally and organically comprise a fair amount of technical knowledge. If the curriculum is made to run along the lines of modern physics it will be necessary to incorporate in it a considerable amount of technical information.

The method of physics must thus devote its attention to the problem of perfecting the curriculum, as well as to demonstrating the increasing importance of physics in the general school program.

If it is recognized that physics is of high importance as a polytechnical and a general educational subject, it must be allocated a sufficient amount of time in the school program, a more extensive curriculum, and improved equipment.

It is impossible to impart sound and useful knowledge in physics without the existence of definite objective conditions, and the method of physics must be quite clear on this point, without trying to find all sorts of far-fetched methods of studying physics without equipment, without good textbooks, and without sufficient time. Neither should the emphasis in the study of physics be placed on studies outside the classroom.

In my article "The Teaching of Physics and the Problems of Technology" which was published in "Fizika v shkole" (School Physics), No. 2, 1956, I stated some considerations as to the improvements which it is desirable to introduce in the curriculum. I will not repeat myself here, but only indicate the most important subjects, whether included in the curriculum or not, which require further elaboration. It is a topical problem for the method of physics to find the best teaching procedure for them.

Let me mention three points in the meachnics course:

a) Methods for the study of the law of conservation of momentum, and of impact phenomena;

b) A discussion on the question of introducing in the secondary-school course a more extensive study of "a body rotating about an axis," of the moment of inertia, and of the application of the second law to the case of rotation;

c) A consideration of the study of oscillatory motion (the oscillation of a point), in association with the circular motion of a material point.

In the course of molecular physics and heat:

a) A much more extensive study of the physical properties of rigid bodies (molecular structure and strains) with a view to technological applications (the treatment of materials and strength computations).

This subject is becoming more and more topical due to the variety of physical and mechanical properties of the materials employed in industry (a wide assortment of special steels and alloys, the application of plastics having a broad range of physical properties);

b) The study of the basic properties of gases in connection with the work of gases in heat engines; in particular, methods of studying gas transformations in heat engines on the basis of indicator diagrams.

In the electricity course:

a) The study of electrostatics based on the unity of charge and field, and some notions of atomic structure of elementary charged particles;

b) The simplest way of studying the quantitative relationships in electromagnetism and electromagnetic induction. The technical application of this knowledge in the elementary calculation of the output of a motor and a generator;

c) The connection of the knowledge in electricity provided in the physics course with the material presented in the electrical engineering course and in the electrotechnical laboratory;

d) The study of electrical conductivity in the light of modern physical views, specifically in the methods of studying the electrical properties of semiconductors;

e) Familiarization with electronic and ionic instruments and with their applications in technology. It is in particular necessary to consider how to explain to the students the idea of electronic computers and their applications in automation.

In the optics course:

a) The presentation of geometrical optics and of optotechnics (discussion of optical instruments and of the concepts of a ray, a beam, and a wave);

b) The presentation of photometry as related to the spectral composition of radiation (the laws of blackbody radiation);

c) The study of spectra as related to molecular and atomic sources of radiation and of modern procedures of spectral analysis;

d) The presentation of the quantum ideas in modern physics.

In the course on atomic structure:

a) The connection of the ideas presented in the chemistry and physics courses on the structure and nature of atoms;

b) The presentation of the radioactive properties of atoms, and experimental techniques in connection with the eventual use of radioactive isotopes in school;

c) Methods for the presentation of atomic power and of the uses of nuclear fuel as an energy source;

d) The presentation of the applications of radioactive isotopes (tagged atoms) in medicine and biology and for the monitoring and automation of production processes.

We have dwelt on a number of points on which elaboration is specifically called for under the new conditions obtaining in the teaching of physics.

Possibly they do not adequately represent the primary tasks of polytechnical training in the domain of physics. The list of these points could certainly be extended and their priority is still subject to discussion.

VII. SCIENTIFIC INQUIRY INTO THE METHOD OF PHYSICS

We have tried to review and evaluate, within limits, the present position of the scientific method of physics, and to indicate the immediate problems with which this pedagogical science is confronted. We have been more concerned with the content of the problems rather than the methods of solving them. This is natural, for we are more interested in the specific aspect of these problems, i.e., the fact that they involve the teaching of physics rather than any other subject.

It is doubtful that the techniques of investigation in the method of physics will greatly differ from the investigation of teaching chemistry or biology.

However, the great importance of experiments in teaching physics constitutes the main difference of the method of physics from that of other subjects. In the method of physics it is even possible to notice a widespread trend to solve all methodological questions "by proceeding from experiment". The search for new or improved versions of old experiments is placed at the forefront of all methodological construction.

It goes without saying that this trend has its virtues, but it cannot be deemed the only possible one. Clearly, the formulation of new methodological problems arises also from the new tasks of the school, from new developments in physics itself, or from its application (e.g., radar), and also from teaching practice.

Teaching practice and its generalization is one of the basic aspects of any systematic study.

Considerations of pedagogic, didactic and psychological order are another aspect.

It seems to us that the best solution for inquiring into the method of physics resides in the joint cooperation of these two fundamental aspects.

Let us dwell once again on a problem which has been considered on various occasions — the problem of experimental teaching in the domain of the method of physics.

Of course, it is impossible to deny that experimental teaching is important, but it would be incorrect to place this technique of inquiry into methodology on the footing of an exclusive and comprehensive method of research.

During his prolonged (more than 25 years) work in scientific research institutes the author often observed pedagogic experiments and conducted many of them.

The most extensive pedagogic experiment with which the author was closely associated was conducted by S. I. Ivanov. This work was concerned with experimental teaching from the 6th to the 9th grades and was carried out on a large scale not only in Moscow schools but in schools in outlying districts. This work proved methodologically quite enlightening both for Ivanov and all those participating in or becoming acquainted with the experiment, though it failed to answer a large number of methodological questions. It is a great pity that the results of this work were not fully published for a number of reasons not relevant to the merit of the work itself. But even those who read all the reports were not quite convinced of the scientific reliability of the conclusions that could be drawn.

In the discussion of teaching experiments the time factor always causes confusion. Given its rate of evolution, the Soviet school never fails to get ahead of any teaching experiment in every form and method worked out mainly by experimental psychology (control classes, comparative quantitative evaluation, etc.).

Of course, we must not deny that experimental teaching is quite valuable as a way of checking the generalizations drawn from teaching experience and based on considerations deriving from physics, from the pedagogic and psychological sciences, and also from personal teaching experience.

We must look for the simplest experimental procedures that can be speedily carried out, and we must pay more attention to the qualitative rather than the quantitative aspect of methodological investigations.

The use of numerical indices is, for the greater part, artificial.

A properly conceived methodological solution should yield a distinctly positive result in a practical test; if, however, the teaching experiment gives a result 5 or 10% above the control experiment, and the "measurers" themselves are up to 90% unreliable, then the results cannot be credited with any validity. We must be aware of the intrinsic significance of social phenomena in the sphere of pedagogy, schools and training, when we ourselves interfere with the process and do not observe it from the viewpoint of a detached observer. We can never secure such mass uniformity in teaching and make it so impersonal as to rely on statistics exclusively.

The most can be derived from a properly performed and accurately observed teaching process provided by the investigator personally or by his assistants who share his views.

This is also, to a certain extent, a natural teaching experiment.

The main thing which hinders the development of the method of physics is the lack of continuity in the work; the failure to draw generalizations or accurate and rigorous conclusions from each work which would allow for scientific development, i.e., the chain reaction on which we have already commented before.

We can only begin something new after a thoroughgoing study of the work which has been carried out earlier.

The careful analysis of the recent history of the method of physics can be the starting point for new syntheses.

There are many reasons for such a discontinuity. We must rectify the neglect of earlier work. We must especially undertake to analyze and summarize all the past work, to evaluate this work carefully and impartially, and only then proceed with new investigations. Anything new can only be built on the basis of what has gone before. Until this is done the method of physics cannot come into its own as a pedagogic science.

We are confident that it is possible to realize this objective.

V. F. YUS'KOVICH

THE DEVELOPMENT OF RATIONAL THOUGHT IN STUDENTS DURING THE TEACHING OF PHYSICS IN SECONDARY SCHOOL

1. Developing the reasoning ability of students as one of the functions of the secondary school physics course

A knowledge of physical principles is necessary, both for the general education of young boys and girls and for their polytechnical training. In the course of polytechnical training the students are acquainted with the physical principles involved in production processes, and with the construction and operation of engines, machines, transmission mechanisms, and automatic devices. Together with knowledge in these fields, the students also acquire a practical working proficiency.

Among the general-educational tasks of the secondary school, of prime importance are the formation of a dialectical-materialistic outlook and the education of the students in the spirit of Soviet patriotism and proletarian internationalism.

For a gradual formation of a scientific outlook in the students it is necessary to interpret in a scientifically correct manner physical phenomena and concepts, to explain the most important laws, and to base the phenomena and laws on general theoretical principles. It is also necessary to exhibit the material nature of physical phenomena. An important aspect of the scientific outlook formed in the students is the certainty of being able to predict the course and outcome of a given set of phenomena, proceeding from a knowledge of the laws of nature. This implies that man is capable of gaining knowledge of the external world on the basis of scientific theory and technological advances. The study of physics conveys to the students the significant features of the modern world.

The effective solution of the above problems implies imparting to the students a definite body of knowledge. But knowledge alone is not enough to enable the young builders to cope with the problems with which they will eventually be confronted in practice. An important task of the school in general and of the physics course in particular is to teach the students how to acquire new knowledge. It is necessary to foster in the students the spirit of inquiry. The acquisition of knowledge has to be accompanied by the intensive stimulation of an active, creative approach of the students to the solution of various theoretical and practical problems. It is at the same time necessary to strive for the development of logical, dialectical-materialistic thought, and of physical and scientific-technical reasoning.

The practical application of the acquired knowledge involves the ability of employing analysis and synthesis, of breaking down complex processes into their constituents, of establishing the causal linkage between phenomena, drawing inferences, constructing rigorous proofs, etc. That is to say, it is necessary to develop in the students the "technique of reasoning," without which the knowledge is liable to become a dead weight.

The importance of developing the reasoning ability of students and of man in general has been pointed out by many scientists, writers, and pedagogues. Let us quote a few statements on the subject.

"Nothing is to be taught on the basis of authority alone, but everything has to be taught on the basis of proof, through the use of the external senses and reason," wrote Jan Komensky.*

The scientific way of thinking requires that "no proposition must be ascerted unless it has been proved by already known truths," stated B. Pascal.**

The noted physicist Paul Langevin maintained that "our modern education is still much too prescriptive: all too often things are l e a r n e d but not understood."†

V. G. Belinskii attributed importance to precise definitions because, in his opinion, the definition brings out the essence of the concept, unlike the mere enumeration of some of its characteristics.

The development of logical reasoning was also deemed important by K. D. Ushinskii. He stipulated that the studies in any school subject should be made to promote the development of logical thinking in the students. Every textbook, according to him, should constitute in its own way a textbook of logic.

These statements which variously underscore the importance of developing logical reasoning stand in sharp contrast to the view expressed by the founder of American pragmatism, William James. He claims, in his "Pluralistic Universe", that he found himself compelled to abandon logic and squarely renounce its findings; he labels reality, if not as irrationally, at least as nonrationally constituted.

The necessity of devoting particular attention to developing the reasoning ability of the students is also due to the fact that logic is not studied at present as a special school subject.

For instance, the special inquiry conducted by N. G. Kushnov in the Leningrad schools led him to the conclusion that the level of factual knowledge of the students in the control class was high. Things were different, however, where logical operations were concerned. On the basis of his investigations, Kushnov writes: "Consequently, the presence of factual knowledge is not enough to ensure the ability of operating with the knowledge in a logically correct manner. This proves that systematically directed work is necessary in the training of the students to reason."††

* Komensky, J. Izbrannye pedagogicheskie sochineniya (Selected Pedagogical Writings), Vol. 1, p. 146. 1906.
** Pascal, Blaise. Oeuvres complètes, Vol. 3, p. 164. Paris. 1903.
† Langevin, P. La Pensée, No. 12, p. 45. 1946.
†† Kushnov, N.G. Vospitanie myshleniya uchashchikhsya v protsesse nachal'nogo povtornogo izucheniya materiala (Training Students to Reason During the Study and Review of the Material). — In: Sbornik Voprosy vospitaniya myshleniya v protsesse obucheniya, p. 119. APN RSFSR. 1949.

The above-mentioned deficiency in the physical knowledge of students is quite prevalent. It stems from the fact that physics teachers pay insufficient attention to the development of reasoning in the course of learning. Admittedly, this is not the only shortcoming in the knowledge of students in physics. Owing to an insufficient amount of practical work, involving the measurement of various physical quantities, the students of many schools fail to develop the necessary sensory-motor abilities and skills. The latter cannot be developed without the proper system of practical courses.

Other defects which remain to be eliminated are formalism and dogmatism in the knowledge of students. It cannot be doubted that the development of the cognitive capacities of the students in the study of physics will have a beneficial effect in overcoming these deficiencies, and in properly assimilating the principles of physics in general.

2. Discussion of the development of logical thought in the textbook and methodological literature

The deficiencies in the way logical thought is developed in the study of physics are due to the fact that the importance of this problem in the method of teaching the school subject is not fully appreciated.

The explanatory note to the 1955 school physics curriculum sets before the teacher such tasks as polytechnical training and the formation of a scientific outlook. But the explanatory note fails to mention the necessity of developing the logical thought of the students. As a result of this oversight in the curriculum, this important aspect has been largely neglected both in the textbook and in the methodological literature. Let us consider several examples.

In the "Physics Methodology" of N.V. Kashin, of which the first edition appeared in 1916 and the fourth in 1923, considerable attention is paid to what is known as "formal education." The author notes three specific aspects in this problem. The first involves the development of the capacity to study the external world. Next comes the necessity of developing in the students the methods of inquiry of the human mind. The third aim is to establish general logical, systematizing methods of inquiry. Putting these objectives into practice will sharpen the senses of the students and teach them how to reason out experimental findings. In Kashin's opinion this method of teaching cultivates the ability of observation and develops the capacity to compare phenomena and to seek the relationships linking them together. This also provides the students with a mental discipline. By the application of scientific methods of enquiry the students learn to make use of induction, and to proceed from observations to hypotheses, laws, principles, and theories.

"In this way," concludes Kashin, "we bring the students to the realization that the methods of the exact sciences are primarily methods of precise thinking."*

The above suggestions, coming from an eminent pre-revolutionary and Soviet methodologist of physics, are of topical interest. Unfortunately,

---

* Kashin, N.V. Metodika fiziki (Physics Methodology), p.14, 4th edition. 1924.

however, the author has not concretely worked out these suggestions in his methodology.

The underestimation of formal logic in general and of its importance in general education in particular has led to the result that the development of logical thought in the teaching of the fundamentals of science in the Soviet school has not been adequately treated in the methodology courses. This applies also specifically to the methodology courses on the teaching of physics. It must be noted, however, that all the authors of methodology courses consider it necessary to tackle the problem of developing the reasoning ability of the students.

E. N. Goryachkin* refers to the necessity of providing the students with a correct outlook and with the elements of dialectical-materialistic thought. He points out further that the students have to acquire the habit of reasoning and the capacity for analysis, in order to strip themselves of preconceived notions and acquire a dialectical-materialistic outlook.

P. A. Znamenskii talks about the necessity of developing the reasoning of the students, of teaching them to proceed from experimental data to modern theoretical concepts or to make use of the latter in the explanation of fundamental phenomena and relationships.

The ability of dealing with concepts is not inborn. "The process of teaching involves not only guidance in the acquisition of knowledge, but also in the mental development of the students."**

Znamenskii maintains that the knowledge acquired by the students should become a means of acquring further knowledge. Obviously, this is made possible by the correct development of the reasoning ability of the students.

In his fundamental treatise on the method of physics,† I. I. Sokolov poses the problem of developing the dialectical thought of the students. The author mentions as the basic requirements for solving this problem the stimulation of the students' thinking, and the establishment of the relationships prevailing between natural phenomena. This must be coupled with the correct interpretation of the basic concepts of physics, i.e., motion, force, mass, energy, the equivalence of the various forms of energy, and the reciprocity of mass and energy.

The foregoing brief survey of the views of leading physics methodologists shows that they are all concerned with the problem of developing the reasoning ability of the students. It is true, however, that the various authors conceive of the problem in different ways.

Articles dealing with the problems of the development of reasoning during the teaching of physics are rarely encountered in the periodicals or in collections. It must be noted that the problem has never been placed within a broad context in the literature on the method of physics. Various articles deal with individual aspects of the question. Thus, for instance, L. S. Dmitriev investigated the development of reasoning as connected with the performance of laboratory assignments.††

---

* Goryachkin, E.N. Metodika prepodavaniya fiziki v semiletnei shkole (Methods of Teaching Physics in the Seven-Year School), Vol.1, p.13. — Uchpedgiz, 1948.

** Znamenskii, P.A. Metodika prepodavaniya fiziki (Methods of Teaching Physics), p.10. — Uchpedgiz, 1954.

† Sokolov, I. Metodika prepodavnniya fiziki v srednei shkole (Methods of Teaching Physics in High School). 3rd ed., p.41. — Uchpedgiz, 1951.

†† Dmitriev, L.S. K voprosu o razvitii myshleniya uchashchikhsya v svyazi s provedeniem laboratornykh rabot po fizike (Developing the Reasoning of Students in the Course of Laboratory Work in Physics). — Fizika v shkole, No.6, p.38, 1949.

The article is interesting, though the author does not differentiate clearly between the various problems involved, i.e., the formation of a scientific outlook, logical reasoning, and the development of research techniques in the course of the work. It is correctly pointed out that the knowledge of the students suffers from dogmatism and formalism, and from an inability to analyze problems. The author suggests developing the reasoning of the students on the basis of concrete knowledge, correlating isolated facts, drawing inferences by logical argument, and selecting and systematizing facts for the proof or solution of a theoretical or a practical problem. The article illustrates by several examples of laboratory assignments how the students develop a critical approach towards observable facts, how their reasoning is formed and they become articulate.

Some information on the development of logical reasoning during the study of mechanics in the eighth grade is given in N. G. Kushnov's article, Training of Students in Reasoning during the Study and Review of the Material.*

Only part of the article is actually devoted to physics. The same problem is considered with respect to geometry and literature. The author confined himself to the examination of several specific aspects, viz., how to teach the students to define, classify, and demonstrate. This valuable investigation unfortunately suffers from some considerable defects. The method of studying some questions in mechanics (dynamics) is by no means correct. Mass, for instance, is defined as the ratio between force and acceleration. After establishing this relationship, the teacher in class, as the author of the article, draws the conclusion that the acceleration imparted to a body by a force is directly proportional to the force and is inversely proportional to the mass.**

Now, since no method is given for measuring the mass, independent of Newton's second law, the formulation of the law involves a logical circle. Moreover, the questions proposed for test papers are sometimes of a dubious nature. Such is, for instance, the question: "Which of the physical magnitudes — mass, acceleration, force, velocity, and momentum — do you consider the most important in the subject just studied and why?" It is not surprising that this question caught inconsistent answers. Incidentally, such examples only prove once again that any departure from a correct presentation and development of the material in the physics course tends to hinder the development of logical reasoning.

B. F. Kil'dyushevskii, a physics teacher in Kuibyshev, writes in his article "The Development of Reasoning in the Physics Class" (Razvitie myshleniya na urokakh fiziki)† about the way of developing the mental activity of the students in the physics classes. The author uses some concrete examples to illustrate the importance of attention and interest on the part of the students in the study of physics, and emphasizes the practical value of knowledge in stimulating the active assimilation of the material by the students.

An interesting article by N. F. Borisenko was published in the journal "Sovetskaya pedagogika" (Soviet Pedagogy).†† The article examines the

---

* Loc. cit.
** Ibid., p. 112.
† Kil'dyushevskii, B.F. — In: sbornik" Voprosy prepodavaniya fiziki v shkole," p. 14. — Uchpedgiz, 1954.
†† Borisenko, N.F. K voprosu o razvitii myshleniya uchashchikhsya v protsesse obucheniya (na materiale fiziki) (Developing the Reasoning of the Students in the Course of Teaching (in Physics). — "Sovetskaya pedagogika," No. 7, p. 13. 1953.

problem of developing logical reasoning in the course of teaching physics in a more direct manner than has been done by other authors. Some concrete examples are used to show the possibility of activating and developing the logical reasoning of the students. Particular attention is devoted to the empirical proof of the different propositions in physics by similarity, dissimilarity, and concurrent changes. The article also dwells on the explanation of the phenomena and their classification. An important place is given to the law of conservation and transformation of energy in the study of the relationships between different phenomena. A valuable point is that the author never fails to mention the specific requirements of physics in developing various logical procedures. Borisenko concludes with the observation that the study of physics is conducive to the development of reasoning if the students learn to perform analysis and synthesis, to consider phenomena from various aspects, to establish relationships between phenomena, to elicit their causes, construct proofs, classify, etc. "Otherwise the study of physics will not have any particular effect on the mental capacity of the students, and they will not acquire any new ways and habits of thinking."*

An extensive experimental investigation of the problems of developing the reasoning of students in keeping with psychological requirements was conducted by A.N. Sokolov, involving the solution of problems in physics in the 6th grade. The conclusions derived by the author deserve serious consideration.

Let us also mention A. I. Uemov's article "Developing the Logical Reasoning of Students in the Solution of Physical Problems" (K voprosu o razvitii logicheskogo myshleniya uchashchikhsya pri reshenii fizicheskikh zadach).** The author takes some physical problems and subjects them to a logical analysis. The article demonstrates the application of deduction and induction, gives examples of drawing inferences by analogy, and analyzes some of the logical errors that are committed in the solution of problems.

I. I. Babushkin, who proved himself to be a creative teacher, also devoted attention to the development of the reasoning of students. In his article "The Development of the Reasoning of Students in Class" (O razvitii myshleniya uchashchikhsya na urokakh) he discusses his experience in teaching in the 8th grade the concepts of velocity and acceleration, with the appropriate formulas and graphs. Particular attention is given to the stimulation of active thinking on the part of the pupils, and the formation of correct ideas related to the everyday experience of the students and to observable experiments.

The underlying idea of the article may be subsumed by the following statement: "By linking the immediate perceptions that the students get of the studied phenomena with the corresponding terms, formulas, and graphs, and by displaying the latter again in experiments, the teacher enables the pupils to tackle consciously the study of the subject, and to understand that the terms, definitions, formulas and graphs are merely a concise way of expressing the relationships between real phenomena."†

Apart from the above-mentioned works, individual aspects of the development of logical reasoning are also treated in a number of methodological manuals. In the journal "Fizika v shkole" (School Physics) this problem is discussed in relation to the definition of physical concepts.††

---

* Loc. cit., p.22.

** Uemov, A.I. Fizika v shkole, No.2, p.37. 1956.

† Babushkin, I.I. In: sbornik" O povyshenii soznatel'nosti uchashchikhsya v obuchenii," p.56, APN RSFSR. 1957.

†† Zemskii, V.A. Opredelenie fizicheskikh velichin v uchebnikakh srednei shkoly (The Definition of Physical Magnitudes in Secondary-School Textbooks). — Fizika v shkole, No.3, p.32. 1955.

This brief survey of the literature on the development of reasoning during the study of physics in secondary school makes it plain that the problem has been inadequately worked out. Some important aspects of the development of reasoning are not even broached in the current literature. Insufficient attention is given to the development of reasoning during the presentation of new material in class.

In the same context, many important subjects, concepts, and laws in the physics course have not been properly analyzed. It should also be pointed out that the development of logical reasoning is often confused with dialectical-materialistic thinking and the formation of a scientific outlook.

D. D. Galanin quite correctly wrote that "the importance of modern physics in the development of reasoning still awaits proper interpretation by pedagogical science."*

Physics teachers and methodologists must keep in view the requirements of the students of the given age and grade. What is involved is the development of their reasoning in the study of a specific physical material. Obviously, this problem cannot be effectively solved without the participation of the student.

In this respect quite a number of problems require elaboration, viz., the way of defining concepts, methods of proving basic scientific statements, the demonstration of laws, etc. In addition, use should be made in all possible ways of the aspects of the physics course which most effectively promote the development of dialectical-materialistic thinking in the students. It is indubitable that only the joint influence of the teacher in class and of the textbook, together with independent work on the part of the student, will achieve the correct development of the reasoning ability of the students.

3. Some general considerations on the development of the reasoning ability of students

Physics is one of the sciences of nature; it deals with the general and simple forms of motion of matter.

Matter, in the form of substance and field and their changes, and independently of human consciousness, exists objectively. Man can become acquainted with the properties of matter, study the phenomena taking place within it, and fit these phenomena into established laws, only through using his power of reasoning. The various physical bodies and phenomena in nature affect man's senses and produce perceptions. The individual properties of the material world are perceived in the consciousness by this elementary psychological process.

In reality, man with the help of his senses and brain can simultaneously perceive not only the individual properties of objects and phenomena, but also create a composite picture in which the various percepts are integrated. The activity of the brain which is caused only by the direct effects of objects and phenomena of the outer world on man's senses, was called by I.P. Pavlov the first signaling system. He defined this system as higher nervous activity.

The first signaling system plays a great part in man's perception of the world in general, and in teaching children in particular, when studying natural sciences and physics. By producing or altering external stimuli, a teacher influences the nervous system of the pupils and produces temporary associations in their minds. Repeated influences strengthen these

---

* Galanin, D.D. Sovetskaya pedagogika, No. 10, p. 26. 1944.

temporary associations and create conditioned reflexes. The intensity and stability of the reflexes are determined by the intensity of the stimulus, the number of nerve centers in the cortex of the large hemispheres that are stimulated, the duration of the stimulus and its frequency of occurrence. Verbalization, i.e., human speech, also affects the large hemispheres of the brain. Pavlov wrote: "Our percepts and concepts concerning the surrounding world are our first concrete signals of reality, whereas speech, and primarily the kinesthetic (i. e., motor —V.Yu.) stimuli reaching the cortex from the vocal organs are the second signals, i.e., the signals of signals. They constitute an abstraction from reality and allow inference, which is our s p e c i- fically human faculty of ratiocination. At first this thought process creates a universal empiricism, and ultimately science, i. e., the tool of man's higher orientation in the surrounding world and in himself.*

Man's thought thus emerges as a function of the influences of the environment on the brain. Thought is a particular form of motion of highly organized material. The process called "reasoning," involves images drawn from reality. It is connected with the presence and formation of concepts. The general properties of things, phenomena, and their mutual relationship are reflected in thought. Psychology asserts that representation involves the visual form of an object, while a concept is the idea of the relevant properties of the object or groups of related objects or phenomena. Consequently, it is not the external aspect or relationship of the objects and phenomena that are embodied in concepts, but it is rather the nature of the phenomena, and their interrelation that are represented.

V. I. Lenin said: "Cognition is a reflection of nature by man. It is not a straightforward, integral reflection, but an operation involving a number of abstractions, formulations, formation of concepts and laws..."**

These basic propositions provide the means of tackling the problem of the development of logical thought during the study of the principles of science in general, and of physics in particular.

Logic is the science of the patterns of thought and the laws which connect ideas in reasoning. Throughout the history of mankind, definite patterns of language and thought were produced, as a result of the long experience of social, industrial, technical, and scientific activities of people. These ubiquitous patterns of thought represent in essence the regularities prevailing in the external world, as reflected in the human mind. Since the laws governing the behavior of nature and of matter are the same everywhere on earth, the laws of thought must be the same for everybody. It is clear that the results of scientific endeavor, and any scientific way of thinking must necessarily comply with the precepts of logic. In the course of history of mankind, four laws of logical thought have been established. Aristotle formulated the law of identity, the law of noncontradiction and law of the excluded middle. The German philosopher Leibnitz (17th century) formulated the law of sufficient reason.

If our thinking is correct and consistent, it ought to conform with the laws of logic and its requirements. It is worth mentioning Engels' views on the subject. In "Anti-Düring," Engels writes that all theoretical thinking is regulated by the inviolable fact that "our subjective thought and the objective world obey the same laws and therefore they cannot contradict one another in their results; they must be in agreement with each other."

* Pavlov, I.P. Izbrannye trudy po fiziologii vysshei nervnoi deyatel'nosti (Collected Works on the Physiology of the Higher Nervous System), p.197. — Uchpedgiz. 1950.
** Lenin, V.I. Filosofskie tetradi (Philosophical Notebooks), p.176.– Partiinoe izdatel'stvo. 1936.

The same idea was expressed by Lenin in his "Philosophical Notebooks." He adduces Hegel's statement to the effect that "every science consists of applied logic, to the extent that it (the science) casts its subject matter into thought and concept."

These ideas are pertinent to the teaching of physics. Sometimes the importance of logical prinicples in the study of physics are not fully appreciated. For example, it is considered possible to divorce logic from the definition of concepts or the demonstration of laws. The fact is overlooked that it is impossible to grasp the significance of a concept or a law isolated from the basic principles of logical thought.

Books on the method of physics often fail to draw the distinction between logical and dialectical thought and between formal and dialectical logic. Concepts such as dialectical-materialistic reasoning and the dialectical-materialistic philosophy are also confused. This loose way of thinking produces further confusion when studying basic problems in physics. The following statement exemplifies the use of such indistinct terminology: "If students knew physics well, all the rest would fit into place."

There is no need to dwell at length on the fallacy of this view. Any violation of the basic principles of elementary (or formal) logic inevitably induces some kind of error of a physical nature. For example, the circular definition of physical quantities detracts from the precision of knowledge.

When dealing in the following with the specific aspects of teaching physics in school, we shall first of all consider the requirements of elementary logic. The teacher must pay attention to the development of logical thought of the students when mastering physical knowledge. If the students are unable to analyze, synthesize, and correlate concrete facts, phenomena, and objects, they cannot be expected to understand properly what these facts involve and to perceive deeper dialectical relationships. Of course, dialectical logic has to be applied in the elucidation of complex processes.

To conclude this section we shall briefly consider the problems of the development of reasoning and the formation of an outlook.

Correct scientific thought fits into patterns determined by logic, and is inseparably linked with language. The laws of thought retain, in broad terms, their validity for all people throughout all ages. Now, outlook bears mainly on the make-up of the contemporary picture of the world, determined by one or a number of sciences. This picture obviously changes with the development of the sciences and of technology. At different stages in the evolution of mankind, the scientific outlook has greatly varied. For example, the geocentric world system of ancient times and the Middle Ages completely differs from the modern heliocentric system.

Newtonian mechanics differs from the mechanics based on the theory of relativity. The class attitude and consciousness of the scientific investigator inevitably has some effect when constructing a scientific picture of the world and interpreting the problems of physics and its discoveries. For example, it is known that discoveries in the field of physics in the last few decades have caused a "physics crisis" in Western science. It has been held that these discoveries led to the "disappearance of matter." In actual fact, the recent discoveries in physics bear out the basic propositions of the dialectical-materialistic outlook.

Therefore, it is necessary to avoid confusion or identification of the rules of thought of universal validity with the system of views on the world which cannot hold good for all periods, peoples, and classes.

Such confusion may cause undue complications in the training and education of students.

4. The function of physics in developing the logical and dialectical reasoning of the students

The basic question is, in which way can the teaching of physics help develop the reasoning of the students? It would be incorrect to solve this problem by including in the study of physics logical concepts and laws, analysis of terms, etc. Such a method of approach would not secure a knowledge of logic and would not make the students proficient in logical reasoning. At the same time it would inject into physics "alien elements," whose study requires methods which are not proper to physics. Thus, what is involved is not the study of the logic in the physics course, but rather the study of physics in keeping with logical principles. This means that neither the content of physics nor its methods of study should be subjected to any changes. At the same time strict adherence to logical principles in the study of the various subjects in physics will be a powerful tool for the acquisition of more profound and conscious knowledge. Accordingly, the reasoning of the students is developed by uncovering the logic inherent in the principles of physics.

In actual practice, the reasoning of the students is developed by guiding their thought from the observation of concrete reality to abstraction and from it to practical application. If no appeal is made to the living impressions of the students and if these are not supported by simple descriptive experiments, their abstract thinking will lack substance and no logical procedures will help fill that gap.

The realm of phenomena studied by physics is fairly extensive. It includes the various mechanical motions of massive bodies, the thermal motion of molecules and atoms, the various motions of electrically charged particles, the motion of photons, etc. Physical phenomena also involve the interaction of material particles with fields — gravitational, electrical, magnetic, and electromagnetic. These forms of motion and interaction vary in complexity. The knowledge of the various forms of motion of matter gradually developed in the course of evolution of physics. Mechanics developed first, more or less independently, and then the theories of heat, electricity, magnetism, and light. Connections were discovered later among these. After the law of conservation and transformation of energy was established, these connections became a controlling factor in the further development of science.

It is impossible to learn physics at present without consistently studying the individual, qualitatively different forms of motion of matter. This division of the school subject of physics into main sections represents one of the major features of its logical structure. Now, in nature and production the mentioned physical phenomena, and often also chemical and biological processes, occur in integrated, undifferentiated form. In science complex phenomena are analyzed, and classified and grouped according to their major characteristics. This is how one subject is divided into different topics of study.

During the introductory lesson in the 8th grade the subject of physics is analyzed and divided into sections, and this already constitutes an important logical operation which is useful in developing the reasoning of the students.

The study of the different forms of motion is closely connected with the formation of a large number of concepts. Owing to the general validity and broad range of application of physical concepts, their definition and correlation is an effective means in developing the reasoning of the students. This holds particularly for a number of general concepts, such as mass, force, energy, work, field.

An important place in physics is occupied by various relationships and laws. Their validity is demonstrated mainly by means of experiment. The educational value of every experiment resides primarily in the fact that it presents the students with the incontrovertible logic of the cause-and-effect relationships prevailing between natural phenomena. K.D.Ushinskii wrote that every experiment trains one's logical ability and that some physical and chemical experiments are more effective in developing correct syllogistic thinking and acuity of observation, than hundreds of exercises written on logical categories.*

It may be added that physical experiments establish relationships between phenomena and also provide the means of expressing them quantitatively. However, the study of the school subject is not enough for this purpose. The combination of the experimental method with the quantitative evaluation of the experimental findings is one of the important characteristics of the physics course.

However important the above mentioned factors may be in the developmnet of the reasoning of the students, they still do not completely resolve the problem. Extremely important in this respect is also the study of some physical hypotheses and theories. This study is accompanied by logical procedures, such as generalization, induction, and deduction.

The only valid way of developing logical reasoning is the scientific, and accordingly logical, explanation of physical phenomena, concepts, relationships and laws, and theories. Obviously, this is methodologically achieved by the teacher's exposition in class, and by the subsequent review of the material. Experience shows that considerable possibilities for developing the reasoning of the students are offered by laboratory assignments and by the solution of problems.

Physical science and the school physics course abound in instances which make it possible to display to the students the intimate relationships linking different phenomena, to illustrate how quantitative changes become qualitative, to show the mutual exclusivity of contradictions in nature, and to establish from the historical development of science and technology their intimate interconnection and the possibility of gradually getting to know nature. These highlights of physical knowledge furnish the students with the elements of dialectical-materialistic thinking. It is possible to do this in all cases involving changes in the form of motion, transformations of matter and energy, and the development of scientific hypotheses and theories. Still, training in dialectical-materialistic thinking in no way obviates the necessity of developing logical reasoning. This twofold purpose is achieved at one and the same time in the senior classes.

The subject matter of the physics course and its methods of study provide many possibilities for developing the reasoning of the students. But in order to exploit them fully the teacher must constantly bend his efforts to encourage the active thinking of the students in general. In order to teach

---

* Ushinskii, K.D. Sochineniya (Collected Works), Vol.2, p.226. —Izdatel'stvo APN RSFSR. 1948.

the students to reason, they have to be stimulated into thinking purposefully. In this respect a great error is committed by the teacher who strives to present the class with ready-made material and does not move the students to perform independent mental work.

One of the physics classes in the 9th grade of a school in Kiev was devoted to the study of rotary motion. The teacher discussed the instruments for measuring the number of revolutions, translatory and rotary motion of bodies, the rotation of the earth and of a flywheel, the motions of a carriage and cutting tool, and angular and linear velocities.

The only question posed to the students at the end of the exposition was:
"Are there any questions ?"
There were no questions.
"If there are no questions, write down your home assignment..."

This procedure is followed all too often in the classroom. But if it is only the teacher who speaks, the independent thought of the students is not made to work. It is left idle. Without excercise there is no development. The right attitude is adopted by the physics teachers B. V. Kil'dyushevskii, N. F. Borisenko, I. I. Babushkin, and some others; they present the students with a planned set of questions and problems, designed to stimulate the active thinking of the students and mobilize their capacities in finding an answer or a solution.

Whatever the grade or the subject studied, it is always possible to draw up a set of questions making clear the principle involved. Such questions are particularly necessary in the sixth to eighth grades. In many cases they help to assimilate the subject matter better in the senior grades.

The questions intended for the students depend on the composition of the class and on the content of the material to be studied. They are quite often accompanied by experiments. In other cases these are simply problem-questions. They sometimes form a logical chain, which draw the students towards the required answer. In almost all cases the problem-questions will provide a stimulus to active thinking on the part of the students.

N. F. Borisenko considers it helpful to give the students such questions as: Why does the water rise after the piston [in a pump] in spite of its weight ? Would a pump work in an airless space ? Why does the mercury not run out of the tube in Torricelli's experiment ? What makes a balloon rise up into the air ? Why does wood float on water, while steel sinks ? How can a ball-bearing be tightly fitted onto a shaft? How can we find what current flows in a circuit (direct or alternating) ?

Each of these and similar questions can be correctly answered only by giving the problem careful consideration and consciously applying one's knowledge.

The teacher of the Kiev school we mentioned before could have also easily given his pupils food for thought. To do this he should have put to the class a series of questions, instead of giving a uniform presentation, almost a lecture, of the material. Such questions could be: What instances of translatory and rotary motion do you know ? What is the angular velocity of rotation of the earth about its axis ? Do points on the surface of the earth in Moscow and Kiev have the same linear velocity of rotation ? What is the difference between the motion of the carriage with the cutting tool and the worked part in a lathe ? What motions do the drill bit and the worked part perform ? (In such questions it should be ascertained that the students are familar with such objects as carriage, cutting tool, drill bit.) What is the relationship between angular and linear velocity ?

Of course, posing these questions, listening to the various answers given by the students, and critically evaluating them as a whole, requires more time than just lecturing on them. But teaching is to be evaluated not on the basis of "time economy", but by the results achieved. The results are represented by the extent and quality of the derived knowledge and its effect on the thinking capacities of the students.

The value of leading questions and heuristic discussions, especially in the the junior grades, is also determined by the fact that they enable the students to proceed from particular instances to general conclusions. They teach the students to employ inductive procedure in gaining knowledge of reality. This method becomes particularly valuable when the discussions are accompanied by experiments.

5. The study of physical phenomena

Physics deals with a great variety of phenomena and processes. These are first consistently studied in the sixth grade, and systematized in the eighth grade. Before defining the subject matter of physics, the teacher examines instances of various phenomena, and performs some experiments. The question then poses itself as to what is common to such phenomena as the falling of bodies to the earth, the expansion of bodies on being heated, the attraction of a magnet on steel, the incandescence of the filament of an electrical light bulb, etc.

The correct answer to this question requires extensive logical processing of examples and experiments. The students can simply be told that all these are instances of physical phenomena. But this kind of dogmatic assertion has no cognitive value. The teacher must therefore choose another way of tackling such cases. First of all the students must be moved to think over the examples and experiments which they have been presented. Let us say that they are to c o m p a r e them, so as to find what may be c o m m o n to the various phenomena. One point that the examples given may have in common is the fact that matter does not change its composition. This distinctive trait of physical phenomena may be made more vivid by contrasting them with phenomena in which the composition of matter is altered. To this end it is possible to use the burning of a candle or the release of carbon dioxide by the action of vinegar on drinking soda. It is then necessary to give a verbal definition of physical phenomena as such phenomena in which the composition of matter does not change. This definition can be used till the end of the 10th grade. Only when studying radioactivity the definition of physical phenomena will have to be supplemented.

At this point some remarks are in order on the necessity of verbal formulation of ideas. Thinking processes and their outcome are both intimately connected with verbalization. Ideas are produced and developed only on the basis of words and propositions. Hence the importance of definitions, which complete the study of any given subject. These definitions are valuable for the development of reasoning only provided they are consistent with the requirements of physics and logic.

In the 6th grade the students have already studied the basics of some sciences: arithmetic, language, history, geography, and botany. On starting the study of physics, the students want to know — what is this new science about? The answer to this question involves the application of analysis and synthesis. Thus, the phenomena occurring in nature are classified by their common characteristics. Some are studied by geography, others by natural science, still others by botany, and yet others by physics.

After defining the domain of physical phenomena, the teacher proceeds to analyze these phenomena. The students thus come to understand the way in which the course is divided into sections, viz: mechanics, heat, electricity, and light.

Now, how can the science of physics be defined? Various definitions are given. For instance, Tsinger states that the word "physics" literally means "the science of nature." This definition certainly does not describe the nature of physical science in its modern context. In addition to physics, nature is also studied by astronomy, chemistry, meteorology, geology, crystallography, mineralogy, botany, zoology, etc.

The standard textbooks for the 6th and 8th grades do not give a definition of the science of physics, which the students are beginning to study. This is a wrong attitude. The definition of physics lays the foundation for the analytic-synthetic way of thinking which runs all along the study of physics.

In his "Teaching Aids"* for the 6th grade, E.N. Goryachkin defines physics as the science of mechanical, sound, thermal, electrical, and light phenomena. The attempt itself to define physics as early as the 6th grade is commendable. However, this definition of physics cannot be considered appropriate, since the subjects of mechanics, heat, electricity, light, and sound were defined a little above as branches of physics. It thus emerges that the definition involves a logical circle. Unfortunately, the experimental textbook of physics for the 6th grade by the same author, published by Uchpedgiz in 1956, does not give any definition of physics at all. At any rate, the proper way would be to define physics in the sixth grade as the science dealing with phenomena in which the composition of matter does not change.

In the second stage of instruction, in the senior grades, it is possible to go further and define physics as the science of the most general and simple forms of motion of matter.

Now, is it possible to give a different definition of physics after two years of study? After all, this may seem to contradict the logical law of identity. This law may be formulated as follows: Any idea within a given argument retains its definite content, whenever and wherever it may occur.**

This law posits d e t e r m i n a c y in thinking. However it does not imply that, given a new stage of development of science, and in the present case of a new stage of teaching, the treatment of phenomena or concepts cannot be refined. The only necessary stipulation is that such refinements should not be arbitrary and a d h o c but should have the proper grounds.

---

* Goryachkin, E.N. Fizika, uchebnye materialy dlya VI klassa (School Physics — Teaching Aids for 6th Grade), part 1, p. 15. — Izdatel'stvo APN RSFSR, 1953.

** Kondakov, N.I. Logika (Logic), p.47. — Uchpedgiz. 1954.

An important part in the study of physics is played by various forms of motion. In the 8th to 9th grades mechanical motion is studied, i.e., rectilinear and curvilinear motion, translatory and rotary motion, uniform and accelerated motion, oscillatory and wave motion. The very fact of drawing a distinction between any type of motion and the motion of physical bodies or particles is quite instructive in the physical and logical sense. In order that the students should obtain a correct conception of each type of motion, they have to be presented with examples, and principally by means of demonstration experiments. Without concrete experience the students cannot get a clear idea of their relevant characteristics. At the same time the study of the different forms of motion must be accompanied by an analysis of the various types, the elaboration of their similarities and differences, and their classification. In the classification of the types of motion it should be pointed out what is taken as a basis for including each in a class. The criterion is the shape of the trajectory of a material point in the case of rectilinear and curvilinear motion, the velocity variations in the case of uniform and accelerated motions, the form of the trajectory of different points of a body in translatory and rotary motion, and the periodic variations in the displacement (or the velocity) of a material point or system of points in the course of time in oscillatory or wave motions. If the teacher wants to produce a well-defined picture of the various types of motion, the students must have a clear idea of the criteria according to which the motions are classified.

What has been said with respect to the analysis and classification of the mechanical types of motion applies also to the various thermal processes, and electrical and light phenomena.

The study of the various types of motion consists in teaching the students how to analyze the complex motions of bodies and machines or their individual parts, and to identify and correlate the different types of motion. It is possible to take as examples the motion of a locomotive, an aeroplane, or a motor-car. These bodies or different parts in them may execute simultaneously or in succession several of the motions described above.

An aeroplane may move in uniform or accelerated motion, in translatory or rotary motion, (in stunt flying), along a curve or a straight line. The different parts of the motion may be subjected to analysis, e.g., the piston or the flywheel in a steam engine, the propeller in an aeroplane. Developing the analytical thinking of the students will enable them to break down complex motions into their constituents, establish the characteristics of each motion, etc. Getting to know the constituent parts of a complex phenomenon proves much simpler than gaining knowledge of the phenomenon in its entirety. But after recognizing the parts separately, it is necessary to proceed with a synthesis of the individual elements of the gained knowledge. It must be noted that the course of physics as represented in the curricula and textbooks gives a certain synthesis of the provided knowledge, though often to an insufficient extent. It is true, however, that this applies more to the technical application of the laws of physics; this will be discussed later on.

The examination of the states of a gas may serve as a good example of analysis and synthesis. In most processes in nature and in heat engineering gases undergo simultaneous changes in volume, pressure, and temperature.

This complex process may be broken down into individual processes. By means of experiment, it is possible to reproduce an isothermal, an isobaric, or an adiabatic process. The characteristic properties of each process are determined from the experiment. The students thus learn to differentiate between gas processes according to whether one or another parameter is constant.

It is not only important to grasp thoroughly these processes and their distinctive characteristics. It is no less important to be also able to synthesize the different processes and to study the state of a gas under arbitrary changes of the parameters.

Developing a proficiency in analysis and synthesis is necessary also in the study of other subjects in the physics course. No matter what the subject studied, it is impossible to do without analysis and synthesis, abstraction and generalization. Let us give a few more examples. The students study individually various thermal processes, viz., the heating and cooling of bodies, changes in the temperature and pressure of the air, the evaporation of water, variations in the air humidity, rainfall, etc. The scientific basis of weather forecasting involves the synthesis of all these elements. The necessity of such a synthesis has to be pointed out to the students, as they often naively assume that the barometer reading is enough to predict changes in the weather.

In the study of electromagnetic induction, in the 7th grade and especially in the 10th grade, the students successively learn about different cases of production of an inductive emf. The latter is produced when the current is switched on and off in the primary coil, when a magnet is moved in and out of the secondary coil, when the coils are brought closer or moved apart, and when the current is varied in the primary coil. However this kind of itemized knowledge is not enough. In this case it is important to make a c o r r e l a t i o n in order to find the basic cause for the appearance of the inductive emf. The question that arises is: what do all the cases described above possess in common ? By abstracting themselves from incidental, irrelevant details, the teacher and the students arrive at the unique cause common to all the above effects. The cause consists in the variation of the magnetic flux running through the secondary coil.

In the section on optics (10th grade) the reflection and refraction of light are usually studied separately. Less attention is devoted to the absorption of light. Now a particularly instructive instance is the general case, when part of the light energy is reflected, part is transmitted by the medium, and part is absorbed. It is desirable to demonstrate this general case by experimental means. The students will thus be able to ascertain once more the value of analytical knowledge and of its synthetic application.

The examples considered do not, of course, cover all the possibilities of developing logical thought in the study of physical phenomena. They merely show that the assimilation of any physical phenomena has to fit into a logical pattern. Only the conscious application of such a logical procedure can promote the development of the students' reasoning. Obviously, rigour and consistency must prevail throughout this cognitive process; thus it must yield conclusive results. A powerful means for achieving this purpose are examples drawn from real life and demonstration experiments.

6. The study of physical concepts

Only by working out a consistent body of physical concepts is it possible to achieve a sound understanding of the phenomena and laws of physics and effectively apply the subject in practice. Of particular theoretical and practical importance are physical quantities. Their importance derives from the fact that physical phenomena and many of the properties of matter are defined quantitatively. The great majority of physical laws display a quantitative relationship between phenomena. Unlike mathematics, which, by Engels' definition, is the science dealing with magnitudes and rests on the concept of magnitude, physics is concerned also with the qualitative aspects of objects, phenomena, processes, and relationships. Every physical quantity embodies in it both the quantitative and qualitative aspects of phenomena or property of matter. This synthesis of the quantitative and qualitative is one of the dominant characteristics of the concepts involving physical magnitudes.

Methods for the teaching of physical concepts, e.g., force, mass, work, energy, current strength, potential, etc., have been worked out and tested in practice. There is no need to deal with the whole complexity of the process of formation of these concepts. We shall concentrate our attention on some specific aspects of this process, which are particularly relevant to the development of logical thought in the students.

The formation of physical concepts in the mind of the students involves displaying the qualitative and quantitative characteristics of the physical quantities, ending with their verbal definition and conscious application in practice.

There is not, nor can there be, a uniform way for defining physical concepts. Definitions vary with the different characteristics of the phenomena or property of matter involved. It is therefore possible to speak only about the most typical ways of solving the pedagogical problem.

In order to enlist the attention of the class and rouse interest in the concepts to be studied, the teacher must first devise ways and means of presenting the concepts. In some instances this can be done by examples taken from everyday life and supplemented by classroom experiments; in some others by an appropriately formulated problem or problem-question to which a solution has to be found. In this way the students are made to realize that a new characteristic of phenomena or bodies is to be learned.

Next the various aspects of the new characteristic or property under study are examined. **The elaboration of the qualitative characteristics** of physical phenomena or the property of matter constitutes an important stage in the formation of concepts.

Any physical magnitude is related in various ways with other magnitudes. In the course of teaching, this relationship can be established by experimental means. It is precisely this relationship which provides a quantitative description of the concepts being studied. Thus a **quantitative dependence** is set up between the new concept and the other, already known concepts.

After the physical magnitude has been qualitatively and quantitatively characterized, it is given a **verbal definition**; the concept in question is thus unambiguously incorporated into a system with other concepts. This, more than other stages in the formation of concepts, involves abstract thinking and plays an important part in its development.

The stages of concept formation considered above do not yet suffice to produce a conscious, sound knowledge of the studied physical magnitude. Only repeated application of the concept in practice can firmly fix it in the students' mind. But this still does not complete the formation of a concept. In the other stages of instruction, when new subjects are studied, the knowledge of the physical magnitude becomes more profound, in content and extent. This is supported by such concrete examples as the formation of the concepts of velocity, acceleration, mass, force, work, energy, field, etc.

It is evident from the foregoing that the formation of physical concepts runs parallel with the development of both logical and dialectical-materialistic thought.

Let us consider the formation of the concept of velocity.

The students have an idea of what velocity means even before they go to school. This initial notion is developed in elementary school, where they solve many problems in arithmetic. The concept is formed further in the physics classes at the end of the 6th grade. Then, in the 8th grade, some new characteristics of the concept are worked out, i.e., the direction of velocity (vector representation), instantaneous velocity.

The elaboration of the concept of velocity begins with a comparison of the motion of various machines and bodies — a bicycle, a motor-car, a train, an aeroplane, an artificial satellite, etc. Given the same kind of trajectory, for instance a straight line, the motion of all these bodies differs in a significant way, namely, each traverses a different distance during the same time. No calculations are required to appreciate this fact. The point may be brought home by demonstration experiments involving the motion of a trolley on a table top, the fall of a ball in air and in water, etc. After that it becomes necessary to find the relationship between the distance traversed and the time. In doing this, of course, one first introduces the mean velocity, then the velocity of uniform motion, and, lastly, the instantaneous velocity.

The concept of velocity must be given precise formulation when the students are first presented with it (the definition will be discussed further on). Subsequently many problems have to be solved, to show the importance of this concept for practical purposes. This will also give the students a chance to get used to the concept. The study then proceeds to linear and angular velocity in rotary motion, and later the rates of various processes are defined, for instance, the rate of change of a magnetic flux, of radioactive decay, etc.*

The students first become acquainted with the concept of mass in the 6th and 7th grades. The study is confined here to some individual examples. First a distinction is drawn between the concept of mass and that of weight. The concept of mass must be applied to the study of some subjects in the theory of heat. At this stage mass is measured only by means of weighing on an arm balance. At this point there is neither the possibility nor the need to make the students conceive of mass as the measure of gravitation or inertia; mass is merely taken to be the measure of the quantity of matter. In the 8th grade, when studying Newton's second law, the students learn

---

* [The same word denotes in Russian both velocity and rate (of change). Thus, to avoid ambiguity, velocity is sometimes referred to as the rate of motion.]

to regard mass as the measure of inertia; finally, only after the law of universal gravitation has been studied is it possible to treat mass as the measure of gravitation. In this way mass is made to appear both as the measure of inertia and the measure of gravitation. Both measures serve to describe the quantity of matter, which exhibits the properties of inertia and of gravitation. The historical fact that these two measures of matter have proved equal need not be used to gloss over the different ways in which the mass of a body is defined.

Making the students appreciate all these facts necessitates a long process of observing phenomena and comparing them, eliciting the features they have in common and disregarding incidental details, making generalizations, and defining concepts. All these steps furnish material for the development of the logical reasoning of the students in the study of concepts.

The formation and definition of concepts must not be seen as a short-term procedure. The process is quite involved. Concepts reflect the properties of objects and of phenomena in nature and production. It often proves impossible to encompass all at once the whole intricacy of many concepts. This means that the definitions of concepts can and should be altered, developed, and refined.

Physics teachers and methodologists often underestimate the problem of defining physical concepts. Some maintain that definitions have to be straightforward and concise, that definitions are a guide to action, etc. Others assert that the definition of physical magnitudes cannot be brought into conformance with logical principles, requiring that the defined concept be qualified in the generic and the specific.

This diversity of opinions on such an important question shows that concept definitions have not yet been adequately worked out in the methodology of physics. At the same time there is no doubt as to the importance of this question for teaching.

It is first of all necessary to dispose of the notion that physical concepts can be presented and defined in any way which is at variance with logical principles. The patterns and laws of thought are universal and any science must comply with them, as must physics and the methodology of physics. As for trying to keep definitions simple and concise, this is quite a creditable aim. Lenin wrote: "Human thought is 'economical' only to the extent that it reflects correctly objective reality, and the criterion for this is practice, experiment, industry."* In another place Lenin points out that although concise definitions are convenient, they are not adequate. Further on he gives his well-known detailed definition of imperialism.

Obviously, a definition must serve as a guide to action. But the definition of a concept is justified in action when it gives a correct representation of the properties of phenomena and objects. It is also incorrect to underestimate the value of definitions in cognition and in teaching.

As Lenin observes in his "Filosofskie tetradi" (Philosophical Notebooks), human concepts are subjective when taken in isolation, but they are objective when taken as a whole, within the source of cognition. Thus, such physical concepts as velocity, acceleration, force, etc., are not introduced merely

---

* Lenin, V.I. Sochineniya (Collected Works), Vol. 14, p. 157.

as a matter of convenience or as arbitrary measuring operations. These concepts reflect the intrinsic properties of objectively existing phenomena or objects. The scientific concepts which have been tested in practice are real in that they give a true representation of reality. These considerations must be kept in mind when the definition of physical magnitudes is tackled.

Some views occur in the physical and methodological literature, bearing on the definition of physical quantities, which deserve at least a brief discussion. In the textbooks of secondary-school physics a method is adopted of defining physical magnitudes by means of ratios between quantities. For instance, velocity is defined as follows: "The velocity of motion is the quantity determined by the ratio between the path and the time during which this path is traversed." The same pattern is followed in the definition of all the other physical magnitudes studied in the 8th, 9th, and 10th grades. Now how correct is this definition, and others similar to it, as far as logic is concerned ? As it turns out, not at all. The definition just considered fails to give the required generic concept, which is not a "quantity" or just any magnitude, but a physical "vector magnitude."

The specific characteristic says nothing about what a given magnitude is, but only how it is measured. While it is true that the way in which magnitudes are measured in physics is quite important, it would be a mistake to reduce physical concepts essentially to measurement. Some physics textbooks, methodology courses, or articles furnish instances of other definitions of physical magnitudes. Thus, again in the definition of velocity, we have: "The term velocity applies to the ratio between the path and the time during which this path is traversed." This definition is not satisfactory, either, from the standpoint of logic. It fails to provide a generic concept, while the specific criterion is reduced to a mathematical ratio between magnitudes. It therefore does not disclose in a clear, logically correct manner the intrinsic nature of velocity.

Some methodologists have tried to consider the "ratio" in this formulation as a generic concept, and take the remainder as the expression of the specific criterion. However, this way of treating the definition does not stand up to criticism.

There occurs still another method of defining physical magnitudes. According to it, velocity is defined as being the path traversed by a moving point in unit time. This definition was upheld, for instance, in a recently published article.* However, this definition is also unsatisfactory as far as logic is concerned, since it gives the path as the generic concept for velocity, which is manifestly incorrect. On the other hand, if the whole statement is considered as a specific characteristic, there is no generic concept in the definition.

The methodological literature in physics lacks model definitions of physical concepts which could correctly develop the logical reasoning of the students.

Now, how can the above problem be most rationally resolved ? Let us point out first of all that there is quite a number of concepts of a general nature which cannot be defined at all.

---

* Zemskii, V.A. Opredelenie fizicheskikh velichin v uchebnike srednei shkoly (The Definition of Physical Magnitudes in the Secondary-School Textbook). — "Fizika v shkole," No. 3, p. 32. 1955.

Some of these are concepts of category. They represent the most general properties of objects and phenomena in nature. Foremost among such concepts are matter, motion, space, and time. When it is stated that "matter is an objective reality, existing outside and independently of consciousness," or that "motion is an inseparable property of matter," or "space and time are forms of existence of matter," such expressions are not to be taken as definitions of the concepts involved. They are nothing more than explanations.

There are also some concepts in physics which do not lend themselves to rigorous logical defintion. Such are: sound, tone, loudness, light, color, warm, cool, hot, red, green, etc. Any attempt to define these concepts yields either a tautology or a circular definition. These concepts are gradually formed in the human mind, as early as childhood, on the basis of sensory experience. To make them clear in the course of teaching does not call for a logical definition, only sensory perception, by experiment demonstrating the phenomena.

O. D. Khvol'son classified as undefinable physical magnitudes, along with extension and duration, also pressure and velocity of uniform motion. The concepts of pressure and velocity can, however, hardly be referred to as categories.

We may draw the attention of physics teachers to the special importance that O.D.Khvol'son attributes to the definition of physical magnitudes. "A definition," he writes, "means the precise formulation of the term it involves, and the only way to do this is by stating the dependence of the defined magnitude on something else already known, i.e., something previously given a precise definition."* Further on he stipulates extreme care and precision in the definition of magnitudes. This implies that the definition should be unambiguous and complete, i.e., "It should comprise everything that might serve as a distinctive characteristic of the magnitude under consideration."**

With respect to teaching, as we have already mentioned, the definition of a concept does not constitute the first step in the process of learning. It comes at the end of some familiarization with the qualitative and quantitative aspects of the physical magnitude. Neither does the definition terminate the development of a concept.

The methodology of physics must strive to define physical concepts only in keeping with the requirements of logical reasoning. "Definitions by the generic and the specific have become firmly established in the every-day affairs of humanity, in school and research procedures."†

Now what would be the cognate generic concepts for physical quantities? Two types may be mentioned: vector and scalar physical quantities. Most of the physical concepts studied in secondary-school can be referred to either of these types. Vectors represent: velocity, acceleration, force, impulse, momentum, pressure, electric field strength, magnetic field strength and induction.

Scalars represent: mass, density, work, energy, output, quantity of heat, temperature, specific heat, potential, emf, resistance, voltage, current intensity, inductance, electric capacitance, luminous flux, illuminance.

---

* Khvol'son, O.D. Kurs fiziki (A Physics Course), Vol.1, p.36. 1933.
** Ibid.
† Popov, N.P. Opredelenie ponyatii (The Definition of Concepts), p.22. Leningrad. 1954.

In such terms, a physical magnitude, such as velocity, may be defined as follows: velocity of motion is a vector physical magnitude, describing the change in the path over a unit of time.

The concept of acceleration, say, may be defined as a vector physical magnitude, describing the change in the velocity of a body over a unit of time.

In a like manner, force is a vector physical magnitude, which defines the deformation or the change of velocity of a body.

Let us now consider some examples of definitions of scalar magnitudes.

Mass is a scalar physical magnitude, representing a measure of the quantity of matter and inertia.

Energy is a scalar physical magnitude, representing a measure of motion during its transformation.

Work is a scalar physical magnitude, representing the measure of the energy converted from one kind into another.

Output is a scalar physical magnitude, equal to the work done in a unit of time.

The potential of an electrical field is a scalar physical magnitude, representing the measure of the energy at any given point of the field with respect to infinity or the earth's surface.

As we said before, the general concept of matter is not defined. However, the forms of matter, i.e., substance and field, can be defined.

Substance is a form of matter consisting of elementary particles, viz., molecules, atoms, protons, neutrons, electrons, possessing a rest-mass.

A field (electrical and magnetic) is a form of matter which manifests itself by its effect on electrical charges or magnetized bodies. The fact that the electromagnetic field is a real entity is attested by the finite velocity of propagation of electromagnetic waves.

We see as a result that the definition of many physical concepts can be made in full agreement with the requirements of both physics and logical reasoning. These definitions, which only expose the most salient features of the given concept, usually also give rise to the appropriate methods of measurement. Of course, the methods of measurement, as well as the techniques and units to be adopted, are subject to further study.

It is obvious that this method of definition can be employed only from the 8th grade on. The students have to become acquainted first with such concepts as magnitude, physical quantity, vector, scalar, and this is to be done in the introductory classes. A vector magnitude may be defined as a physical magnitude having dimensionality and direction. A scalar magnitude is defined as a physical magnitude possessing only dimensionality (e.g., volume, area).

The concepts of vector and scalar magnitudes might at first be rather restricted in extent and content. But the subsequent study of each new physical concept will supplement and enlarge them.

By following this procedure in the definition of the principal physical magnitudes, the physics teacher breaks a vicious circle. He will encounter no difficulty in the definition of each new concept, since the definitions will not run counter to the requirements of logic.

It is also about time to put an end to a pointless debate as to the advantages of one method of definition over another, without taking into account the principles of logic. Protracted discussions and debates as to the way of defining physical concepts have up till now been fruitless. Neither is this situation liable to change as long as any particular method of definition is sought without strict adherence to logical principles. Also in the method of physics teaching definitions must be in full agreement with the requirements of physics and logic. It is possible that the definitions proposed here should have to be amended. But any improvements must be sought in keeping with logical requirements, and not in avoiding them.

It is also necessary to give up the idea of defining all physical concepts in a uniform way. It is clear from the foregoing that the definition of scalar and vector magnitudes should be intrinsically different. There is, in addition, another kind of physical magnitudes studied in secondary school, namely, the various coefficients. These may be divided into two groups. The first group comprises the dimensional coefficients of proportionality, e.g., the coefficients of linear and volume expansion, the coefficients of thermal pressure, the mechanical equivalent of heat, the thermal equivalent of work, the gravitational constant, the coefficient of elasticity and Young's modulus, the gas constant. These coefficients are associated with certain laws and define the relationship between various magnitudes in the adopted system of measurement. With respect to these physical magnitudes it is meaningless to ask whether they are scalars or vectors. These quantities may however be larger or smaller, and have definite qualitative characteristics. These coefficients of proportionality may therefore be defined as physical magnitudes expressing definite ratios. Thus, for instance, the mechanical equivalent of heat is a constant physical magnitude which causes the same heating effect in a body as a unit quantity of heat imparted to it.

We may mention, incidentally, that in the textbooks for the 7th and 9th grades this concept is defined somewhat differently: "The amount of work equivalent to a unit quantity of heat is called the mechanical equivalent of heat." This definition sounds tautological and is thus not acceptable. Also, the definition makes no mention of a generic concept.

The fourth group of magnitudes* is that of abstract coefficients. These express ratios between two magnitudes of the same kind. To this type belong the coefficient of friction, the coefficient of efficiency, the electrical permeability, the magnetic permeability, the index of refraction. All these magnitudes have only a quantitative characteristic. In this respect they are similar to the number π which represents the ratio of the circumference of a circle to its diameter. These concepts may be defined as magnitudes or numbers equal to the ratio between corresponding magnitudes of the same kind.

Certain cases may occur where the concept involved does not call for any definition. Some such concepts were already mentioned before. In the 6th to 7th grades most of the concepts have to be given descriptively, explained by examples, experiments, etc. Thus there is not much point in trying to define what electricity is in the 7th grade, or even in the 10th grade. In such cases it is necessary to appeal to explanations, and draw up experiments to compare, correlate, differentiate, etc.

Here, for instance, is how the writer D. Granin explains to his readers through the words of his hero, what electricity is: "Electricity is the ideal energy, of almost limitless versatility; it can be stored up, and transmitted

---

* [In addition to vectors, scalars, and dimensional coefficients.]

over thousands of kilometers; it can give light and heat, melt metals, and cut things; it can make wheels turn, it can blast holes, it can talk, it can break down matter."* This account of the properties of electricity and its uses could give the students a much more vivid idea than any kind of definition. It is sometimes worthwhile to make use of such literary descriptions to explain the nature of certain concepts.

We may conclude the consideration of the definition of physical magnitudes with the stipulation that the definitions should be clear and free from tautology. Furthermore, they must be well-balanced, in the sense that the defining concept should cover as much as the defined concept. It is also necessary to avoid circularity in the definitions. This kind of error is committed, for instance, in the physics textbook for the 7th grade. The voltage across a section of a circuit is first defined there (§97) in terms of the electrical energy carried over that section when a unit quantity of electricity passes through. When the students come to the calculation of the work and energy of current (§97), the definition of the voltage is brought in and the work is calculated as the product of the quantity of electricity and the voltage. It thus turns out that the voltage is defined by means of the energy of the current, and the energy is defined by means of the voltage. The definition of the voltage also contains another mistake. One unknown is defined in terms of another unknown, the voltage being defined by means of the electrical energy, which remains subsequently to be defined itself.

In order to avoid circularity in the definition of the voltage and the work, the concept of the voltage must be defined in the 7th grade independently of the work. This may be done by starting with the study of electrical current, current sources, and Ohm's law. This is also desirable for the reason that the manner of presentation of the voltage given in the textbook is difficult to follow by 7th-grade students. We feel that there is also a mistake in the definition of work and energy in the textbook of the 8th grade. Thus, the concept of work is introduced at first as the product of the force by the path (§61). Obviously, this is not yet a definition of work, but only a way of calculating it. In §70, energy is defined as the ability of a body or a system of bodies to perform work. The students are then given a full definition of work, as the measure of conversion of energy (§77). The result is a circular definition, since energy is first defined by means of the work, and then work is defined by means of the energy.

This logical error can be avoided by introducing the concept of energy separately, independently of the concept of work, as the measure of motion of bodies being converted into another form. Logical consistency and scientific accuracy are especially necessary in the study of energy and work, considering the great importance of these concepts in the physics course.

The formerly current treatment of work as the product of the force by the path has been lately supplemented in many methodological works and textbooks interpreting work as the measure of conversion of energy from one form into another. At the same time energy itself is still considered as the available amount of work of a body or system of bodies.

This new way of treating work has been made necessary by the fact that its interpretation as the product of the force by the path is confined only to the domain of mechanical phenomena. But the concept of work can be used in a broader sense. It is possible to speak of the work of an electrical

---

*Granin, D. Iskateli (The Seekers — a novel), p. 116. 1955.

current, of radiation, etc. The concept of work in its enlarged sense covers all the cases of conversion of one form of energy into another in various processes.

However, it is incorrect both from the logical and the scientific standpoint to regard work as a measure of converted energy, and energy as the amount of available work.

Furthermore, work is not the only measure of energy conversion. Another measure is the amount of heat. In this way it would seem that energy does not represent only the amount of available work since, by analogy, it would be possible to describe energy as the available amount of heat. This, however, is obviously false. A body does not contain any available heat. Now how can we escape this dilemma? The new interpretation of work cannot be given up, for the reasons already mentioned above. There appears to be only one issue, and that is to change the manner and order of presentation of the concepts of energy and work.

We may outline the order of presentation of these subjects in the 8th grade. The students are first acquainted through examples and experiments with the different forms of motion and their transformations. Experiments suitable for this purpose are the motion of a ball in a curved trough and of a ball suspended on a string or on a spring. Such experiments clearly show the variations and transformation of motion.

Next the attention is drawn to the conservation of motion. The above given experiments already lay the foundation for this. The point may be further driven home by means of Maxwell's pendulum experiment. To analyze these motions, the concept of energy is introduced as a measure of the motion. The concept is expanded later to include the measure of interaction of bodies.

At a certain point the question arises as to the value of the energy. In order to show how the energy is measured it is necessary to make an experiment, in which some equal loads are lifted to a given height. Suppose five weights, each weighing 1 kg, are raised to a height of 1 m. It is possible to lift five times one kilogram, or once five kilograms. Other combinations are also possible. The students find out that in all these cases the product of the weight by the height remains the same. It is thus concluded that the value of the energy that is conserved is measured in the present case by the product of the force by the displacement and is expressed in kgm. The concepts of potential and kinetic energy are then introduced. After that it is explained how energy is converted in various technical devices, such as a hammer, a pile driver, a hydroturbine, a wind motor, etc., and the law of conservation and conversion of energy in mechanical processes is formulated.

Every power machine is designed for the specific purpose of converting one form of energy into another. It thus becomes necessary to enlarge the concept of work. This concept is in fact introduced in order to evaluate the converted energy. The latter is obviously also measured in kgm.

Later on (in the 9th grade) the students are acquainted with another measure of energy conversion, viz., quantity of heat.

At that point the students are already familiar with the concept of work, from the 6th grade, and know how to calculate it in terms of the product of the force by the path, expressed in kgm. This treatment of work becomes a special case when the expanded version has been learned.

Needless to say, the commission of logical errors in the study of any physical subject can only hamper the development of logical and physical thought alike. When defining any individual concept, the physics teacher must strive to achieve both scientific accuracy and logical consistency.

7. The study of physical laws and theories

Several score different laws are studied in the secondary-school physics course. These are intrinsically significant relationships linking phenomena together.

Each time that a law is discovered in physical science the study of nature rises to a new, sometimes considerably higher, level. Such was the case, for instance, with the laws of dynamics, Newton's law of universal gravitation, and the law of conservation and transformation of energy established through the efforts of Descartes, Lomonosov, Meyer, Joule, Lenz, and Helmholtz.

The theory of electricity was given a strong impetus by the discovery of the law of the electromagnetic induction by Faraday's and Maxwell's far-reaching investigations, which led to the development of electrical engineering, the invention of the radio, and the achievements of modern electronics.

An extremely important part in science is played by physical theories, which serve to correlate and systematize a variety of phenomena and individual facts. Such are the molecular-kinetic theory, electron theory, the theory of atomic and nuclear structure, the theory of light, etc.

The study of the different laws and theories in the physics course should run parallel with the development not only of the scientific outlook but also the reasoning of the students. Unfortunately, all too often such is not the case. This is due, among others, to the fact that the students are not taught to draw the proper distinction between facts and their implications, to differentiate between laws and hypotheses and theories. A major defect noted in many schools in the teaching of physics is a dogmatic approach to the presentation of laws and theories. The latter are given as the self-evident results of research.

We had the occasion of observing this way of studying the laws of reflection of light in a 10th-grade class. The teacher says: "You have already studied the laws of reflection of light in the 7th grade." He then proceeds to present the question as if the students had never heard of the subject. Not a single question was posed to the students. The teacher drew on the blackboard the path of the incident and reflected ray, and marked with the chalk the angles of incidence and of reflection.

"If we measure these two angles in practice, we find that they are equal."

The experimental setup prepared for the lesson was never used. None of the students had the slightest doubt as to the truth of the formulated laws.

Proceeding from these laws, which are very simply demonstrated, the teacher shows how to construct an image in a plane mirror. Again it is the teacher who does all the drawing on the blackboard.

"Is it possible to measure the distances of a luminous point and of its image from the mirror?"

"It is. These distances are equal," answered a student.

That was all. No need was felt to give any grounds for this answer, for instance, to give a mathematical proof of the quality of the distances. Apparently the students had not been taught in the physics classes to reason logically, to prove or demonstrate laws and other significant statements. This is the result of the indiscriminate use of dogmatism in teaching.

The physics textbooks also have their share of dogmatic assertions. Much too often they include statements such as: "scientists have found," "it has been scientifically established," "it has been experimentally proved," "it can be shown." Thus, for instance, in the physics textbook for the 7th grade we have the statement: "It has been established in science that energy cannot vanish or be created" (p. 151); in the textbook for the 10th grade we read: "It has been established by modern science that the electrical field is one of the forms of matter" (p. 7). Many of the formulas in the textbooks are given without derivation.

One of the tasks of education is to impart to the students a body of well-founded knowledge. A most important means of validation in physics, in so far as it is an experimental science, is the experiment. When physical laws are studied extensive use must be made of experimental proof. Proving a law means inferring the existence of a relationship between phenomena or properties of a body from experimental results. In proofs of this kind the teacher applies the inductive method (incomplete induction). Thus, what is found to hold for one or two experiments is generalized as being true for all experiments of the same nature.

Let us consider a possible way of studying Pascal's law with respect to liquids and gases in the 6th grade. In order to make the students appreciate the significance of this law, it is necessary, as usual, to appeal to the everyday experience of the students and to classroom experiment. We proceed from the understanding that it is possible, and indeed methodologically desirable, to study this subject for liquids and gases separately. At this point, however, in order to save time and space, we shall consider this law for liquids and gases jointly.

The students can be given the familiar example of the water pipes used in practice. Many of the students may have read or heard about oil and gas pipelines (for instance, the Saratov-Moscow pipeline extending over more than 900 km). Also, among the many devices employed in the study of the atmosphere and the stratosphere use is made of sounding balloons.

The question arises as to how these contrivances function. What is the physical principle on which they rely?

In order to answer these questions, let us consider some simpler examples and experiments.

Teacher: "Have you ever blown up a paper bag, in order to make it pop?"
Student: "Of course, many times."
Teacher: "What happened when you blew into it?"
Student: "The paper bag swelled out all around."
Teacher: "What happens to a hot-water bottle when water is poured into it?"
Student: "The hot-water bottle also swells out."
Teacher: "Can anyone think of some other similar examples?"
Student: "I have often had to inflate a football."
Another student: "I have seen in a shop how toy balloons are filled up with gas."
Third student: "I have blown soap bubbles in the bathroom."

Teacher: "Well, the examples you gave me are quite good. Now can you tell me why a football, a bubble, a hot-water bottle, or a paper bag swells out on all sides when it is filled up with water or gas? In order to make it easier to answer, let us first perform two experiments."

The teacher then shows the well-known experiment with Pascal's balloon or some other simple experiments.

Teacher: "What is common to the phenomena in these experiments and the previous examples?"

Student: "Both in the examples and in the experiments there were containers which were filled up with water or with air."

Teacher: "True. Have you noticed anything else in which they were similar?"

Student: "The water and the air exerted a pressure on the walls of the containers."

Teacher: "Precisely. Now, what do we know about this pressure?"

Student: "We saw in the experiment with the balloon that the jets of water squirted out from the holes in different directions."

Teacher: "That's correct. What we found today in the class was discovered back in 1653 by the French physicist Pascal. We dealt only with water and air, though. The law that Pascal discovered holds true for all liquids and gases. This law states that the pressure on a liquid or a gas is transmitted in all directions with the same force."

After this the teacher will be able to explain many practical applications of the law. For instance, it will become clear to the students why water can flow from different taps on the same floor.

In the 6th grade the students have to learn the law of Archimedes. This can be achieved in various ways. Let us consider one of the possibilities. The teacher may get various answers to the question whether the students know what keeps boats or ships, submarines, or airships from sinking. Some students might know the reason, others might not. After discussing the answers the teacher may point out that it would be interesting to know the actual reason that keeps such bodies afloat or aloft. To this end it is necessary to study an important law of nature.

Teacher: "Have you ever gone swimming in a river or in the sea and tried to pick up a heavy object from the bottom, a stone, for instance?"

Student: "Yes, certainly."

Teacher: "What did you notice if you took a large stone out of the water into the air?"

Student: "The stone became heavier."

Teacher: "Why is that?"

Student: "I don't know."

Teacher: "Well, actually the explanation is not so simple."

The teacher then proceeds to demonstrate to the class an appropriate experiment, for instance with Archimedes' bucket and water.

Before the cylinder is plunged into the beaker with water the dynamometer pointer reads zero. This is pointed out to the class.

It is also necessary to show that the volume of the cylinder is exactly equal to the capacity of the bucket.

Teacher: "What will happen if the cylinder is plunged into the water?"

Not every student will be able to answer this question correctly. However, it must be asked.

Teacher: "Suppose we try to find out by doing the experiment. . . . What do you see now?"

Student: "The pointer of the dynamometer goes up. . . ."

Teacher: "Why should this happen?"

Student: "Apparently, when the cylinder is plunged in the water it pulls less on the spring."

Teacher: "Correct. Now why does the cylinder pull less on the spring?"

Student: "It is probably pushed up by the water."

Teacher: "What you said is quite right. Now think a minute, what is the force with which the water pushes the cylinder out?"

Student: "We have to measure that."

Teacher: "How?"

Student: "We can pull down on the spring so that the dynamometer pointer returns to its former position."

Teacher: "That is right. Only we shall not pull the spring down, but we shall pour water into the bucket. I will pour and you will watch the pointer. . . . What do you see now?"
Student: "The pointer goes down."
Teacher: "I have filled the bucket up to its brim. Where does the pointer stand?"
Student: "It has come to its former position."
Teacher: "What is the conclusion that can be drawn from this experiment?"

Of course, the students will not be able to give a definitive formulation of the conclusion on their own. It is necessary to make them think, however, while the teacher derives the conclusion. It should be noted than an inexperienced teacher might, after just one experiment with water, definitively formulate the law for **all liquids**. In order to prove the law as it really holds, the same experiment should be performed with other liquids, for instance with a copper sulfate solution, kerosene, saline water. Only in this experimental way, appealing also to observations from everyday life, is it possible to derive the law of Archimedes in a generalized form, for all liquids.

The students are liable to get the idea that the law of Archimedes is somehow dependent on the specific use of a bucket, a cylinder, and a dynamometer. Conclusions of this sort may be effectively forestalled by demonstrating the same results by means of some different arrangments, for instance, with a beaker of water set on a balance, etc.

In order to drive the point home the students may be made to perform some experiments themselves, with other liquids. Repeated exercise in setting up an experiment and tracing out its logic will enable the teacher to inculcate the habit of correct scientific and logical reasoning. The experiments are used as an aid to teaching how to apply **incomplete induction** as a means of validation and proof.

In the following lessons some conclusions are drawn from the law of Archimedes; at that point it can be explained how and why boats and ships float. It is a rather difficult task to teach the students how to apply a general law in the explanation of particular cases. But it is necessary to apply this deductive method of acquiring knowledge as far as possible, especially in the senior grades. The physics teacher must make use of proof by induction in the study of all the basic laws — Pascal's, Archimedes', Ohm's law for part of a circuit, the laws of dynamics, the gas laws, Hooke's law, Ohm's law for a complete circuit, Joule's law, Faraday's laws of electrolysis and electromagnetic induction, Lenz' law, and the laws of illumination and of reflection and refraction of light. Each grade, of course, imposes its own specific requirements on the way in which the laws are studied. In the junior classes an important place is given to heuristic talks. However, the logic of demonstration, and the transition from individual observations and experiments to broad generalizations, must be well established.

We proceed to consider two more examples. Let us examine the way in which Newton's second law is studied, as it is set forth in the textbook of the 8th grade.

The presentation begins with the force of gravity, and it is stated that the weight of a body is related to its mass. After that mass is dogmatically defined as the quantity of matter contained in the body. At the end of that subsection (§48) it is pointed out that when a balance is in equilibrium the mass of the body is equal to the mass of the weight. This means that a corollary of Newton's second law is implicitly used at that point. In fact, the body and the weights that are balanced will be of equal mass only if both are equally accelerated at the earth's surface. If the same experiment is repeated with a very long balance arm laid along a meridian, the equilibrium would be upset. Next, on the basis of experiments, relationships are drawn up between force and acceleration, and acceleration and mass. The second law is formulated and then applied to the case of terrestrial gravitation. There follows a derivation of the formula of the force of gravity (§49), which had been implicitly employed much earlier. In this way the proof of the law turns out to be circular.

Another logical error is also committed in this proof. Mass is defined as a measure of the quantity of matter, and then used in Newton's law as a measure of inertia. The law of identity is thus violated. There is therefore a logical inconsistency in the proof of Newton's second law. In order to avoid the above-mentioned errors, either of two conditions must be fulfilled: 1) mass should be defined, if this is feasible, before studying the second law and independently of it; 2) mass should be defined only on the basis of the second law, without being introduced earlier.

It proves impossible to refer to and define mass as a measure of inertia prior to the second law. Therefore, only the second alternative remains.

The theoretical and practical importance of the laws of dynamics, and particularly Newton's second law, makes it necessary to find the correct way of studying them. As is known in history, the formulation of the laws of dynamics was the outcome of extensive practical experience, and not of any single experiment. When Newton's laws are studied in secondary school, however, it is necessary to work out each law on the basis of a few examples and experiments. Now where the second law is concerned, it should be borne in mind that before the law has been studied the students are already familiar with the concepts of acceleration and force and with the manner and units in which they are measured. At that point the students still know nothing about mass as a measure of inertia.

Newton's second law intrinsically establishes a relationship between acceleration and the force producing it. Viewed in this way, the second law says nothing directly about the relationship between mass and force or between mass and acceleration. These relationships are the principal corollaries of the second law of dynamics.

In order to explain the law a number of examples are first examined.

The following conversation could take place between the teacher and the students.

Teacher: "Many of you have had the occasion to throw a stone or to kick a football. What would you do in order to throw the stone farther or to make the ball reach a greater distance?"

Student: "In order to make the stone fly farther, it is necessary to throw it harder. In order to make the ball roll farther, it must be given a stronger kick."

Teacher: "What kind of motion do the ball and the stone perform?"

Student: "Both the stone and the ball perform a uniformly decelerated motion. The larger the distance they have to cover, the higher the velocity they have to be given."

Teacher: "That is correct. But it is not quite clear how the force comes into it."

Student: "I don't know exactly how to connect this with the force. But, practically, a large force has to be applied to make the stone or ball reach a great distance."

Teacher: "Your observations are right. Let us now try and analyze another example. An empty boxcar is standing on rails. One workman can make it move with difficulty. Two workmen do it more easily, and three of them even more so. How would you explain this?"

Student: "According to the law of inertia the boxcar remains at rest as long as there is no force acting on it. The workman has to apply a force in order to start the boxcar moving."

Teacher: "That's right, but it is not all."

Another student: "The applied force causes the boxcar to go into motion. The larger the force the quicker the boxcar reaches a high velocity."

Third student: "At the beginning the velocity of the boxcar is zero. Then it starts moving. So there has been acceleration. When one workman pushes the boxcar the acceleration is less than when two or three of them are pushing."

Teacher: "Your reasoning and conclusions are correct. But we need to know the exact relationship that holds between the acceleration of a body and the force applied to it. This can be found only by experiment."

An experiment is next performed, with a small ball running down a groove in an inclined plane, under various angles of inclination.

Teacher: "How can we explain what we just saw in the experiments?"

Student: "When the trough is inclined at a small angle the force moving the ball is smaller than at a large angle of inclination. We can see this if we resolve into components the weight of the body lying on the inclined plane."

Another student: "The experiments show that at small angles of inclination the velocity changes at a slower rate than at large angles. This means that in the first case the acceleration is less than in the second."

Teacher: "All right, then what is the relationship between the acceleration of the ball and the force acting on it?"

Student: "From the experiment we can conclude that the ball is given a higher acceleration when a larger force acts on it."

Teacher: "Quite true. Now in order to measure the force we will perform one more experiment."

An experiment is performed with a loaded trolley moving under the action of various forces. The value of each force and the resultant acceleration are written down on the blackboard. The acceleration is calculated from the length of the path and the corresponding time interval, measured with a stop-watch or a metronome.

Teacher: "What can we say about the acceleration of the trolley with the load and the force that was acting in each case?"

Student: "We can see from the experiments that the acceleration of the trolley with the load is directly proportional to the force."

Teacher: "How can we denote this relationship mathematically?"

Student: "If three values of the acceleration turn out to be a_1, a_2 and a_3, and the corresponding values of the force are F_1, F_2 and F_3, we can write: $a_1 : a_2 = F_1 : F_2$ or $a_2 : a_3 = F_2 : F_3$"

Teacher: "In what other way can we write this relationship?"

Student: "We can also write: $\frac{F_1}{a_1} = \frac{F_2}{a_2} = \frac{F_3}{a_3}$."

Teacher: "And to what is equal this ratio of the force to the acceleration in the case of the trolley with the load?"

Student: "This ratio is equal to a constant magnitude for the given body, so:

$$\frac{F_1}{a_1} = \frac{F_2}{a_2} = \frac{F_3}{a_3} = \text{const}.$$

Teacher: "Yes, all this is correct. This constant magnitude, equal to the ratio of the force acting on the trolley with the load and the acceleration imparted to it, defines the inertia of the trolley with the load. The ratio of the force to the acceleration it produces accordingly serves as a measure of the inertia of the trolley. This measure of inertia is called the **mass** of the body and is denoted by m, so that $\frac{F}{a} = m$.

"We have derived this result from our experiments with the trolley. If we were to determine the acceleration and the force in the motion of a stone, a ball, a boxcar, we should find that the result holds good for all these bodies. This result expresses therefore a general law of nature. It was discovered in the seventeenth century by the English physicist Newton. How could we formulate Newton's second law?"

Student: "The ratio of the force acting on a body to the acceleration it produces is equal to the mass of the body."

Teacher: "We have mathematically expressed this in the concise form $\frac{F}{a} = m$.
"This may be more conveniently made to read:

$$F = ma,$$

i.e., the force acting on a given body is equal to the product of its mass by the resultant acceleration."

The way of studying Newton's second law as presented above has the advantage of being consonant with scientific and logical principles. It does not involve any logical inconsistency and is free from circularity. The students proceed from the observation of specific phenomena, gradually work out a quantitative description of the motion and the acting force, and finally arrive at the intrinsic properties of the phenomena, i.e., to the law. This procedure also constitutes an exercise in the application of incomplete induction.

All the foregoing also applies to the study of Ohm's law for a complete circuit. This law is given in the textbook proceeding from a statement of the dependence of the current strength on the voltage and the resistance. Then, when resistance has to be eventually explained, this is done on the basis of Ohm's law.

The same defect is present also in the exposition of Ohm's law for a section of a circuit. Also in this case the dependence of the current strength on the voltage and the resistance is established first of all.

This inconsistency can be avoided if the current strength is established as being dependent on the voltage for any given conductor, while the resistance is defined on the basis of Ohm's law.

The above examples show once again the hazard involved in ignoring the principles of logic in the exposition of the subject matter. The net result is that, instead of being developed, logical reasoning is inhibited.

It is a fact, however, that the secondary-school physics teacher does not always have the possibility of proving a law or a theory experimentally. This applies, for instance, to the law of universal gravitation, or Coulomb's law. Also, modern physical theories have been based on many experiments, e.g., the experiments of Millikan and Ioffe to measure the charge of the electron, or Rutherford's experiments on radioactivity and atomic disintegration, which cannot be performed in school. This does not mean, however, that they have to be given inconsequentially. In such cases it is necessary to discuss the basic principle involved in the experiments, the experimental setup, and the results derived therefrom. In this way the history of science becomes a source of knowledge and provides a basis for it.

Once a law has been proved it becomes an objective fact. Subsequently, when the physics teacher has to prove the validity of any particular point, he must necessarily appeal to the derived laws and use them as a basis for the explanation of further phenomena.

Of particular interest in teaching are those cases when some laws may be derived by deduction from another, more general law. In this way the relationship prevailing between different laws is shown. In this respect the law of conservation and transformation of energy is of special importance. In the course of their physics studies the students have various occasions to apply this law, in various contexts; on each occasion this is done by induction. The students proceed from the consideration of individual examples to more and more comprehensive generalizations. In the 6th grade the law is formulated on the basis of only several examples of mechanical nature. In the 7th grade it is made to cover thermal and electrical phenomena. In the 8th grade, the law of conservation and transformation of energy is already cast into quantitative form. The students then deal with the calculation of the potential and kinetic energy of a body raised above the earth's surface, and some other phenomena are also discussed, e.g., impact, deformations. In the 9th grade the quantitative expression of the energy is used to determine the relationship between work and heat. Unfortunately, at this point the law is not employed to explain some processes taking place in gases, or to explain the changes occurring in the state of aggregation, i.e., fusion and solidification, boiling and condensation. Inadequate use is also made of the law in the study of heat engines. In the 10th grade, the law of conservation and transformation of energy is not applied to any sufficient extent.

It is very important that the students should understand the basic facts and experiments which were used in the history of physics to derive general principles by induction. No less important are also the many deductive consequences stemming from them. The students should be shown that Pascal's laws, the golden rule of mechanics, and the laws of Bernoulli, Joule, and Lenz are particular cases of a more general law.

The study of these and many other subjects in physics provides concrete examples of inferential reasoning. When the students learn the basic physical law of conservation and transformation of energy, as well as other laws, they see that the law develops historically from the knowledge of an isolated fact, through the discovery of what is specific to it, to the derivation of a general principle. F. Engels has shown this in the case of the law of conservation and transformation of energy. It was primarily established that friction is a source of heat. This inference of an individual character was drawn from practical experience of long standing.

At the beginning of the nineteenth century many experimental investigations led scientists to the conclusion that every mechanical motion can be converted into heat through friction. This is an inference of a specific nature.

Soon after it was established that any form of motion can be transformed into any other form of motion. This deduction is a general principle. New discoveries can only add to the content of the law. Engels writes: "A universal law, in which the form and content are of equal generality, is not capable of any further extension. It is an absolute law of nature."*

In the same way it is worthwhile to consider the deductive derivation of Newton's first law — the law of inertia — from the second law, under the assumption that the acceleration of the motion is zero. The law of universal gravitation † gives rise to the fact that there is a force of gravity on the earth, and thus to the laws of motion of falling bodies. The planet Neptune was discovered by means of the law of universal gravitation, on the basis of empirical data on the motion of the planet Uranus. The same inductive-deductive procedure subsequently led to the discovery of the planet Pluto.

The law of conservation of matter is used to derive in a straightforward manner the equation of continuity of fluid flow.

It is a good idea to show, not as a theoretical proof but only as the calculation of a particular case from a more general principle, how Ohm's law is derived from Joule's law, or how the gas laws are derived from the equation of state of gases. Theoretical proofs will not stand up in these cases,

---

* Engels, F. Dialektika prirody (Dialectics of Nature), pp.178-179. — Gospolitizdat. 1950.

since in the emprirical investigation and derivation of the general laws of Joule or of Clapeyron's equation use is made of other special principles.

We should stress the fact that most physical laws may be expressed in a mathematical form which is fairly simple and within the grasp of the students. The mathematical formulation of any particular law makes it applicable to a broad range of practical cases. For instance, using the law of refraction of light, which involves the index of refraction of the substance, it is possible to explain the course of a light ray in optical instruments, e. g., a projection lantern, a camera, a microscope, a telescope, a spectrograph, and also to calculate the optical power of lenses.

The examples considered above provide many possibilities for the development of the reasoning of the students when studying the laws of physics.

In order to make use of these possibilities the teacher must observe certain requirements. In most cases the students learn physical laws through the empirical observation of phenomena and the determination of their causes. By means of incomplete induction it is then possible to establish a general law of nature. In many cases the students' understanding of a law can be improved by a heuristic explanation. In proving laws it is quite important to avoid inconsistencies and logical circularity. The students should be given a clear and precise formulation of every law they study. In order to develop reasoning it is important to connect special principles with a general law (e. g., the conservation of matter, the conservation and transformation of energy, universal gravitation). The law established is further employed in the solution of a number of relevant practical problems. In this way the habit of deductive inference is inculcated. These are the principal aspects of the study of physical laws, viewed as a means of developing the logical reasoning of the students.

Let us review briefly the significance of physical theories in the development of the reasoning of the students. In this case it is of prime importance to make the students appreciate the difference between empirical facts and phenomena, and the hypotheses and theories used to explain these facts and phenomena. In order to make this distinction quite clear, it is necessary to examine consistently and systematically many physical phenomena in the field of heat and electricity, and atomic and nuclear processes.

Empirical facts of this kind are the constant temperature of fusion and solidification, and of boiling. In the same line we have the energy consumed on melting solids or vaporizing liquids. The question that poses itself is how to explain these facts. This is where the molecular-kinetic theory steps in, to provide an answer to this question.

The same thing applies to electrical phenomena. The empirical facts are the electrification of bodies by friction or contact, the heating of wires by the flow or current, the temperature dependence of the resistance of conductors, the electrical conductivity of metals and electrolytes, the lack of conductivity of dielectrics and the low conductivity of semiconductors, the release of substances on electrodes, the production of inductive emf, etc. The students become acquainted with these facts through experiment. When the question arises as to how to explain them, the electron theory is called upon.

One of the typical mistakes noted in the answers of students, even in the senior grades, is that they fail to differentiate between the facts on which theories are based and the theories themselves. This is inadmissible, and may have regrettable consequences. If such a situation occurs in the study of physics, it is only the result of insufficient attention on the part of the teacher to this important aspect.

It is well known from the history of physics that some theories have been superseded by others, while the physical facts have remained the same. The imponderable fluids used to explain electrical and light phenomena have been replaced by the modern electron, wave, and quantum theories. The caloric theory has been superseded by the molecular-kinetic theory of matter.

The conceptions on the ether may serve as another characteristic example. In 1898, the eminent Russian physicist O. D. Khvol'son wrote: "Is not this ether also a metaphysical element that still survives? Will it not share the fate of the five fluids and should we not strive to do away with this remnant of old delusions? To this question we answer: No, a thousand times no!" Further he says: "The existence of the ether is not to be doubted at present any more than the rotation of the earth about its axis and around the sun."*

Not more than seventeen years later, science was enriched by fresh facts and by a new theory — the theory of relativity.

In 1915 Khvol'son stated in one of his lectures: "It must be said that the ether has no longer any use in science, and therefore many eminent scientists now completely deny its existence."**

This example shows how unwarranted it is to confuse facts with their interpretation and the danger in assigning absolute truth to theoretical propositions and conclusions. This, however, is not an argument for abandoning theories altogether. The greatest minds in chemistry and physics have always been firmly in favor of hypotheses and theories in science.

D. I. Mendeleev maintained that "It is better to uphold a hypothesis which may in time prove false, than none at all. Hypotheses provide assistance and guidance in scientific work, i. e., in the search for truth."†

In the tenth grade a good example of the development of hypotheses in a physical theory can be the interpretation of light phenomena. The appearance of the corpuscular and undular hypotheses on the nature of light constituted the first scientific attempts to explain the phenomena and laws of propagation, reflection, and refraction of light. These attempts were based on analogy with the way elastic bodies and sound are bounced off, the propagation of waves on the surface of water, etc. The discovery of defraction, interference, polarization, and double refraction of light and their explanations by means of the undular theory definitively instated the wave theory of the luminiferous ether. In the middle of the nineteenth century it was experimentally proved that light is propagated in water with a lower velocity than in air. This gave the death blow to the corpuscular theory, since, according to this theory, the velocity of light in water should be higher than in air. The evolution of physics eventually led to a new electromagnetic theory of light. None of the two earlier theories turned out to be fully correct. With the discovery of new phenomena, viz., the interaction of atoms with a luminous flux and the emission of light by atoms, the quantum theory of light was created.

---

* Khvol'son, O.D. Pozitivnaya filosofiya i fizika (Positivistic Philosophy and Physics), p. 15. 1898.
** Khvol'son, O.D. Znanie i vera v fizike (Knowledge and Faith in Physics), p. 13. 1916.

† Mendelev, D.I. Sochineniya (Collected Works), Vol. 1, p. 89.

At the end of the 10th grade, after having learned many facts and laws, the students come to a dialectical conception of light phenomena. Light is, simultaneously, both a wave and a quantum electromagnetic process. By reviewing the historical course of development and the contemporary state of ideas on the nature of light the students will be able to evaluate critically the various scientific theories and to consider them as stages towards an increasingly deeper understanding of physical phenomena.

A scientific theory provides the means of unifying into a whole a variety of facts, and makes it possible to systematize and correlate a diversity of phenomena. Theory is a tool for penetrating into the innermost workings of phenomena and thus makes it possible to understand and explain them. By means of theory it is possible to predict the probable development of phenomena under a given set of conditions. On the one hand, theory originates from the correlation of facts; on the other hand, once it has been scientifically established, a theory provides the means of deducing from it significant consequences. Thus, for instance, on the basis of the molecular-kinetic theory it is possible to interpret correctly the temperature and pressure of a gas, to derive the gas laws of Boyle-Mariotte, Gay-Lussac, Charles, and Avogadro, to calculate the velocity of gas molecules , etc.

It should be emphasized that without theories it is impossible to gain an insight into the great variety of physical phenomena and laws. This is as true of scientific investigation as of teaching.

"A theory," wrote Academician L. I. Mandel'shtam, "stands in the same relationship to individual laws as the laws to individual phenomena."*

Let us deal with one example in some detail. Consider a phenomenon such as the fusion and solidification of bodies. The students are already acquainted with these processes in the 7th grade. They are shown by means of experiments with ice or naphthalene that the temperature of fusion and solidification is constant. They become familiar with the points of fusion and solidification of some metals and alloys. But the process itself of fusion and solidification, or the mechanism of disruption of a solid crystal and its crystallization, and the constancy of the temperature in these phenomena can be explained only by means of the molecular-kinetic theory.

The teacher demonstrates to the students these processes when they are studied. Now then, the question arises as to how the process of fusion, say, is explained by the molecular-kinetic theory. The students already know this theory in the 9th grade. They are also familiar with the structure of crystalline bodies, in whose lattice are arranged molecules, atoms, or ions.

Furthermore, the students also know the law of conservation and transformation of energy, which holds good for the process of fusion, as well as for any other phenomenon in nature or industry. The use of a known theory and law for elucidating a new phenomenon furnishes the students with fresh physical knowledge. As to the question stated above, the energy inparted to the body is converted into an internal, molecular-potential form of energy. This produces a change in the forces of interaction between the atoms of the crystal lattice. The regular spacing of the particles in the crystal is thus disturbed, and the solid turns into a liquid.

The application of physical theories to the interpretation of various relevant subjects constitutes a good exercise for the logical thinking of the

---

* Mandel'shtam, L.I. Polnoe sobranie sochinenii (Complete Works), Vol. 3, p. 357.

students. Similar exercises are involved in the study of evaporation and condensation, in the explanation of vapor pressure, or the heating of wires by an electrical current, etc.

In order to provide for the effective development of reasoning during the study of physical theories, it is first of all necessary to give the students a clear idea of the phenomena, processes, or laws involved. This is best achieved by observation and experiment. The principles of a physical theory are presented on the basis of a certain number of phenomena. The knowledge of the students improves further on as the theory is applied to the explanation of new phenomena.

This application of theory is accompanied by some deductive inferences. It is quite useful to show the power and importance of scientific theories by using examples on the prediction of new facts and relationships, e. g., electromagnetic waves, light pressure, etc.

8. Developing the elements of dialectical thought in the study of physics

The study of the principles of modern physics, even in an elementary exposition, cannot be achieved while keeping within formal, or elementary, logic alone. Naturally, some aspects in the study of the various concepts, laws or theories will go beyond elementary logic.

Elementary logic, applied to the study of physics, makes for a correct approach to the classification of phenomena, the definition of concepts, in analytic and synthetic thinking, and in drawing up inferences and deductions; it imposes necessary conditions in the proof of different laws, etc. Elementary logic helps establishing simple external relationships between phenomena, between cause and effect. But there exist in nature not only permanent objects, relative rest and equilibrium, or simple relationships between phenomena. There are also various changes, motions, and transformations taking place in it, and there are many subtle relationships between phenomena. Some physical phenomena, forms of motion and matter are transformed into others. Only dialectical reasoning is capable of encompassing and adequately describing these changes. Elementary logic is not enough in order to understand the way in which one kind of energy is transformed into another, how matter passes from one state into another, or how some elementary particles turn into others. The comprehensive study of the complex intrinsically significant relationships prevailing between physical phenomena calls for dialectical logic.

The study of physics, as that of any other science, cannot be divorced from logical reasoning. But something else is true as well. Physics could not possibly be learned strictly in formal-logical terms. Now education calls for the formation of a dialectical-materialistic outlook in the young generation. This objective is in turn connected with the development of the important elements of dialectical thought. The proper training will thus enable the school graduates to avoid falling into errors of a metaphysical, mechanistic, or idealistic nature.

The history of science abounds in examples where even great scientists failed to evaluate correctly new discoveries, because of a metaphysical way of thinking. A case in point, involving an undialectical view on the microcosm, is that of Lord Kelvin, who up till his death (in 1907) regarded atoms as indestructible particles of matter. He never accepted Rutherford's conclusion that radioactivity results from atomic disintegration.

The Soviet school is expected to inculcate and develop the principles of dialectical thought and to ensure that adolescents form a true materialistic outlook. Physics constitutes the basis of all natural sciences and technology, and its study in secondary school necessarily poses the problem of developing the elements of dialectical thought in the students.

To understand how some forms of matter are transformed into others, and the reciprocity of the various kinds of motion, as well as to interpret many physical processes, is possible only by acknowledging the objectively given transition of quantitative changes into qualitative ones, and the intrinsic identity of conflicting forces in nature. In such cases it proves insufficient to make use of the categories, laws, and formulas of elementary logic and it is necessary to appeal to dialectical logic. Excercising its application promotes the development of the elements of dialectical reasoning, and it is only by means of the latter that the variety of phenomena, concepts, and theories can be correctly represented. In this particular aspect we have in mind the senior grades of secondary school. In the junior grades (the 6th and 7th) some basic aspects of logical reasoning are developed.

It should be possible to solve this problem by the same means as are used in tackling the other problems of teaching, i.e., the development of logical reasoning, the formation of a materialistic outlook, and polytechnical training. What is involved, therefore, is bringing to the fore the dialectical nature of the physical phenomena and processes, relationships and laws that are studied, the historical evolution of theories, and the interdependence of human affairs and science.

Dialectical thinking and human creativity are not inborn attributes. They are inculcated in the course of teaching and depend on the way in which it is carried out. In this respect it is certain that the elimination of dogmatism and formalism from teaching practice would be of great value. The students must be consistently taught never to use a proposition without the proper foundation and adequate proof. The application of research techniques in performing some practical laboratory assignments and solving experimental problems will also help developing the creative thinking of the students. For this purpose it is highly valuable to work out the basic design of different mechanisms and machines, to seek new ways of solving problems and setting up experiments, and to give original examples. Efforts should be made to employ all the concrete means and methods of studying the principles of physics in order to develop the creative dialectical thinking of the students.

In developing dialectical thought in the students there is no need to appeal to any artificial similes. Physical phenomena and the relationships between them provide plentiful material for this. Let us consider a few examples.

As early as in the study of mechanics the students have to deal with various motions of bodies. They thus learn the connection between space and time. No matter what bodies or particles may be moving, the motion always takes place in space and time. Grasping this fact is of prime importance for the subsequent understanding of many macro- and microprocesses.

It is important to demonstrate to the students, by means of concrete examples and experiments, the relativity of any motion that is studied or observed. At the same time motion is also absolute, since there is no such thing in nature as a motionless body. It is in this specific sense that motion is an inseparable property of matter. This characteristic of motion must be pointed out. Unless the students understand that, they are liable to regard motion on an equal footing with rest and equilibrium. Any state of rest is actually purely relative. In speaking of mechanical motion, it should be noted that it is the simplest kind of motion. Further on, in the study of thermal, electrical, intra-atomic, and chemical and biological processes, the teacher will have to point out this particular characteristic of mechanical motion. More complex forms of motion of matter in the inanimate and living realm are accompanied by, though not reducible to, mechanical motion.

One of the mechanical phenomena that deserves special attention is the interaction of bodies. It is very important that the students should properly understand the fact that a force is produced only by the reciprocal action of two bodies on each other. Specific manifestations of the interaction of bodies or material particles are the forces of gravity, of friction, of elasticity, of molecular adhesion, and Coulomb and nuclear forces. When studying these phenomena and ascertaining the presence of the corresponding forces, it is necessary to determine their nature, i.e., whether they stem from the interaction of material bodies or of elementary particles. Obviously, it is only gradually that the students will come to appreciate the full extent and implication of the concept of force. Other concepts that are gradually developed in the mind of the students are velocity, mass, energy, etc. This has been already discussed before.

During the study of heat, electricity, light, and atomic structure it becomes possible to display before the students the dialectical nature of physical phenomena. We may mention in this connection such phenomena as heating and cooling, fusion and solidification, evaporation and condensation, the ionization and recombination of molecules, their dissociation and molization, the emission and absorption of energy by atoms, and photoelectric phenomena. In all these phenomena it is fairly easy to show the transition of quantitative changes into qualitative ones. The variation of the temperature or the pressure in some processes, of the concentration of molecules in solutions in some others, of the frequency of radiation in still others, brings about a corresponding qualitative change in one process or another, a transition from one state into another, a change in the resistivity of electrolytes, etc. It is important to point to the concrete aspect of these phenomena and processes. Stated in physical terms, the students must grasp the dialectical nature of the phenomena. A philosophical generalization "on the transition of quantity into quality" may be given only in the last school-year, and then again proceeding from a firmly grounded knowledge of physical processes and laws.

The variability of scientific ideas, hypotheses, and theories can be well displayed in the study of the development of views on the nature of heat, electricity, light, and the structure of matter. A brief historical review of the changes undergone is valuable in developing a dialectical outlook on science, its content and methods of investigation, and on the relativity and inherent limitations of different scientific ideas. This has also been discussed previously.

The course of development of dialectical regularities is well illustrated by the evolution of views on the structure of matter: from the solid, smallest indivisible particle of matter — the atom of Democritus (4th — 3rd century B.C.) — through the interpretation of the atom as a body that cannot be split into two (Maxwell, middle of the 19th century) and Thompson's "smeared" atom, to the planetary atom of Rutherford and Bohr. But this was only the beginning of the probing into the secrets of the atom. The atomic age began with the artificial disintegration of the atomic nucleus, the discovery of the neutron and the positron, and, most important, the practical application of nuclear energy. These are victories of materialistic science over the idealistic conceptions in physics. It is a known fact that many scientists at the end of the nineteenth century and the beginning of the twentieth completely denied the real existence of molecules and atoms.

In the final term of the physics course the students are given the most important facts about elementary particles. Natural and induced radioactivity, the interconversion of elementary particles, the conversion of an electron and positron into energy quanta and vice-versa — these and other facts illustrate well the dialectical nature of microphysical processes. And even though idealistically-minded physicists give to the conversion of an electron and positron into photons the name of "annihilation" of matter, and to the converse process the name "materialization" of energy.

9. Some requirements for the work of the teacher and students

The problem of developing the logical reasoning of the students cannot be effectively resolved without constant attention on the part of the teacher. To be sure, the elements of logical reasoning are always developed to a certain extent when the principles of physics are studied. But in order to enhance the effectiveness of teaching in this respect it is necessary methodically to make use of the wealth of possibilities inherent in the subject matter itself as well as in its methods of study. The complaint is often heard that the students' knowledge of physics is poor, inadequate, or formalistic. This may be attributed to two main causes. There is first of all insufficient independent work done by the students in the laboratory, in observation and in carrying out practical assignments. The level of knowledge is also appreciably affected by the attention that the teacher devotes, or fails to devote, to the logical development of the subject matter. Quite often there is a lack of unity between the concrete perception of physical phenomena by the students and their logical elaboration.

The learning process in teaching is made up of the direct sensations and perceptions of the students and the guiding explanation of the teacher. Either of these factors is determined by the content and the method of study. It may be assumed that many deficiencies in the knowledge of physics are due to the fact that both the teaching and the independent work of the students are done in a purely prescriptive manner. The students often learn the subject matter without trying to interpret and understand it. Understanding, however, can be achieved only by long and systematic exercise.

In order to learn to analyze and synthesize, to apply induction and deduction, it is necessary to make frequent use of these procedures when studying fresh material, when reviewing it, during laboratory work, in the solution of problems, and in the preparation of home assignments.

Of prime importance in the development of logical thought are the way in which the study material is presented and the relevant explanations in the textbook. These are the two principal ways of gaining knowledge, and if they are inadequately treated the development of the reasoning of the students ultimately suffers thereby.

It can be often observed how little of the lesson time is devoted to a logically correct presentation of the subject matter. This does not give the teacher the possibility to lay the proper foundation to the study of the subject matter involved. In such cases the demonstration experiments are superficially carried out. The students are thus not given the opportunity of drawing a conclusion in a logically continuous manner. Often the teacher himself adopts a stereotyped approach to the presentation of the material. The subject of the lesson is first written down on the blackboard. Sometimes the appropriate sections of the textbook are given at that point for home study. The teacher then proceeds to say: "Let us perform an experiment...." No attempt is made to draw the attention of the students to the subject being studied by suitable questions. In this way the students are not impelled to think actively.

There are, of course, many instances where the physics lesson is well planned and conducted. Let us examine one of the lessons given in the 7th grade in the school which we mentioned before in connection with the dogmatic study of the laws of reflection in the tenth grade. The lesson involved the study of magnets and of the magnetic field. The preliminary review took about ten minutes. One student defined with ease a magnetic pole. He was familiar with the fact that cast iron and steel are attracted by a magnet. With a magnet in his hand and by means of a magnetic needle on the table he demonstrated how magnetic poles interact and then formulated the law of interaction of magnets. Another student showed a good knowledge of the units of electrical current watt-second, hectowatt-hour, and kilowatt-hour. He easily drew up the relationship between these units.

The teacher then proceeded with the presentation of the subject "The magnetic field of the earth." He drew the attention of the students right away by asking a question.

"You may have wondered why a magnetic needle always points to the north. Now what is the cause of this phenomenon?" The teacher then told the class about the first guesses that the earth is a great magnet, and that this magnet has two poles. He brought up facts, known to the students from geography, on the poles of the earth and on the meridians. The facts were provided by the students themselves. A large geographical globe stood on the table. The teacher held in his hand a magnetic needle suspended on a horizontal axis. Another needle with little pennants at its ends rested on a pivot. The teacher gave a description of the angles of inclination and of declination and showed these angles to the class by means of the suspended magnetic needle. As it turned out, a bar magnet had been placed inside the globe beforehand. In this way the students were able to visualize the different positions of the magnetic needle at various latitudes. The students were positively holding their breath as they watched the magnetic needle traveling over the globe. The teacher further told the class about the navigation compass, magnetic storms, the Kursk magnetic anomaly, and magnetic prospecting.

The explanation of the new material took about thirty minutes. The students seemed too reluctant to leave when the bell rang at the end of the lesson.

Thus, here we have two grades — the 10th and the 7th — at the same school but with different teachers. The students of the 10th grade found it difficult to learn the material studied in the 7th grade (the reflection of light), while the students of the 7th grade encountered no difficulty with phenomena and concepts which are studied in detail in the 10th grade.

In one class the physics lessons consisted of learning diagrams and the dogmatic assimilation of concepts and laws. The thinking of the students was reduced to passive contemplation. In the other class the students were engaged in active thinking, concretely analyzing the physical phenomena and concepts. In this case the study of physics served as a basis for understanding its practical applications.

Many teachers always find the means of sustaining the attention and interest of their students at a high level. To this end the new material is presented by way of discussions, giving examples and performing experiments that are not included in the textbook; the teachers use quizzes and make the students exchange questions and critically evaluate each others' answers. A leading factor in educating the students to think logically is the logical reasoning of the teacher himself during the lesson, especially when explaining new material. The explanations of the teacher must not only be plain, clear, and descriptive; they must also be accurate and logically consistent. It is particularly important that they should be convincing and lead to conclusive results. It sometimes happens that the teacher has not had the time to demonstrate any experiments, and he then hurries to generalize, draw the required conclusion, and formulate a rule or a law. To be sure, experiment alone is not enough. It is also necessary to drive home the point that whenever the same set of conditions obtains the same phenomena are invariably produced. In this way the students learn the concept of causality in natural phenomena. Now it rarely happens that an experiment performed during the exposition of the material should be repeated later on, during examination or review, despite the fact that this is indispensable for the proper appreciation of the subject and for its logical development. Sound knowledge is difficult to achieve otherwise.

The function of the teacher when presenting new material, and later when reviewing it, is to think out loud with the class. This makes it easier for the students to grasp the principle involved in the subject matter and the logic of its development. The teacher may also find it helpful to make such comments as "let us analyze the experiment or the problem," "let us try to correlate and generalize these examples and experiments," "let us now make use of deduction, to derive corollaries from this law," etc. Obviously, after making such remarks the teacher proceeds, together with the students, to carry out the analysis, synthesis, generalization, deduction, etc.

There is yet another aspect which should be considered. The development of reasoning can be effectively achieved only provided one school subject rests on another, and constitutes a continuation of it. In this context it is indispensable that related or common topics, concepts, and laws should be interpreted in the same way in physics, chemistry, and other subjects. The opposite is not uncommon, though. Gas phenomena, the gas laws, electrolytic processes, and other questions are treated differently for the different subjects. In mathematics, for instance, theories are based on rigorous proof. In the study of mechanics, however, the necessary conditions are not always observed. It thus becomes necessary to adjust constantly the methods of study of many questions common to the various subjects, especially in physics, mathematics, and chemistry.

The rigor with which the teacher has to express himself in class applies to the students as well. The teacher must carefully listen to the answers of the students, and teach the rest of the class to do the same. Any physical or logical errors in an answer are brought up for discussion. The task of the teacher is to teach the students to express their thoughts correctly in form and accurately in content. The language of the students is often cluttered up with incorrect expressions which distort the physical meaning of the phenomena they are discussing. Expressions such as "the body releases heat," "the heat goes over into work," "the work goes over into heat," "the available heat in the body," are common. It must be made clear to the students that only energy of one form can go over into energy of another form. Work and heat are merely two equivalent measures of these transformations.

It is just as erroneous to say that "the force is expended," "the work is consumed." Instead one ought to say that "the force is applied to the body" or "the force acts on the body," "the work is performed" or "the energy is expended."

Any lack of precision in the way of speaking denotes an unclear way of thinking. If a student says "a calorie is when a gram of water is heated up to 1°," then he does not have a clear idea of the units in which heat is measured.

The students cannot answer all questions in the same way. When discussing a phenomenon the students should describe its course of development, perform an experiment, indicate the distinctive characteristics of the phenomenon, and mention the causes that produce it. It is important that the student should give examples drawn from his own observations and from technology.

The discussion should be to the point and built up so that one thought follows from another, and that the conclusions are drawn up with an understanding of cause and effect. For instance, this requirement is not fulfilled in the following answer of a 10th-grade student: "An ammeter has a low resistance because it is connected in series, while a voltmeter is connected in parallel and therefore its resistance is high." The student has not noticed that in the first conclusion the cause and effect have been interchanged. The two parts of the same statement are differently constructed.

In the consideration of concepts it is very important to teach the students not to restrict themselves to a definition only. Whether the question involves pressure, force, mass, work, or energy, it is necessary to have an understanding of the phenomena on the basis of which the given concept is worked out by generalization. The teacher may appraise this understanding by having the students give examples from everyday life, describe experiments, and demonstrate them on simple devices. A definition is, of course, necessary, and it must comply with the requirements of logic.

In connection with this, let us quote K. D. Ushinskii: "Whoever has a clear idea of an object can indicate the class of objects to which it belongs and specify the distinctive characteristics which differentiate this object from other members of its class."*

---

* Ushinskii, K.D. Sobranie sochinenii (Collected Works), Vol.4, p.568.—APN RSFSR. 1948.

The students must be able to interpret physical laws as real relationships prevailing between phenomena or magnitudes. The students often tend to think of laws as something extraneous introduced into the phenomena. The point must be driven home to them that laws are indissolubly connected with the phenomena themselves. The laws are not appended to phenomena, but are inherent in them. The students sometimes misinterpret laws in their answers, to the point of unintelligibility. Cause and effect are often interchanged. Here is the kind of answer that can be given by some 10th-grade students: "Every body remains in a state of rest as long as it does not need an external force to remove it from its rest;" "there is a proportional relationship between mass and acceleration, from which Newton derived his second law." The conditions under which a law was derived are also misrepresented. A seventh-grade student gives the answer that "the ancient scientist Archimedes took a stand and a dynamometer and discovered his law." The principal thing is that a law should be established and proved mainly by experimental means.

It is of considerable importance to make the students appreciate the universal validity of the laws of conservation and transformation of energy, of matter, and of momentum. Some other laws are confined to given sets of conditions, for instance, the laws of Pascal, Boyle-Mariotte, Gay-Lussac, and Ohm. The proper understanding of the meaning and content of the laws can be best checked by the way in which they are applied to the solution of practical problems, or the explanation of the operation of machines. Careful attention should be paid to the answers of the students in this particular respect. The formulation of the law alone, even if it is correct, is not a sufficient criterion by which to judge the knowledge of the students.

In answers to questions relating to physical theories, it should be made sure that the students draw the proper distinction between individual facts and their explanation by means of the theory. The students must be able clearly to represent and interpret different kinds of phenomena, e. g., diffusion, Brownian motion, the states of aggregation. The answers should cover the demonstration of the phenomena involved, and their description. The students must have just as clear an understanding of the fact, say, that the molecular-kinetic theory is based on the assumptions, that: substances consist of molecules and atoms; the molecules are constantly in motion; there are forces of interaction between the molecules. The validity of these assumptions can be tested by experiment and is supported by all the facts known at present.

In the physics course many different technical devices are studied. A clear understanding of the physical principles involved in them and their mode of operation ensures a sound knowledge that is of practical value. The study of the construction of various machines develops the technical outlook of the students and also offers the opportunity of comparing these machines, finding points of similarity and difference in them, and this promotes the development of a logical and technical way of thinking. For instance, all heat engines have much in common; these engines convert the internal energy of the fuel into mechanical energy, and they have heaters and coolers. But there are even more things that differentiate these engines from one another. Pertinent questions will reveal the students' knowledge in this respect. It is important that the students should clearly perceive the physical principle involved in the machine, understand the construction of its basic components, be able to point them out on a model and explain their purpose and mode of operation. However, the understanding of the operation of the machine as a whole and the capacity to work on it are properly developed later on, in the courses of machine construction and the fundamentals of production.

Now, the students are supposed to draw their knowledge of the various principles of physics from the explanations of the teacher and from the textbook. In order to secure a positive effect on the thinking of the students, it is desirable that the general logic of concept formation and the development of the subject matter in the textbook should run along the same lines as the logic of the lesson. It would be impossible to achieve good results in the development of reasoning if these two major means of gaining knowledge find themselves at variance with each other. It was not an idle remark of Ushinskii's that every [physics] textbook must be in its own way a textbook of logic. Every lesson must also be an example of rigorous logical thinking.

In order to assimilate the principles of physics the students have to perform independent work systematically; a considerable portion of this work is done in the classroom, and part of it at home. A particular form of independent work which helps developing the reasoning of the students is the solution of problems. A considerable number of reference works have been written on this method of study.

The solution of problems is also discussed at length in the methodology courses on the teaching of physics. There are also special psychological investigations dealing with the development of reasoning during the solution of problems. This method of study has been most fully worked out with respect to the development of the reasoning of students. There is no need to discuss the solution of problems in detail in the present paper. We would only like to draw the attention of teachers to some aspects of the questions which are of particular importance for the development of the logical reasoning of the students.

The solution of physical problems entails complex mental activity. The students are required to apply their theoretical knowledge to the solution of a given practical problem. This is a process involving the careful analysis of one's knowledge and the selection of the facts suitable for one's purpose. Whatever method the student adopts for the solution of a problem, he necessarily has to break down the problem, determine its physical and technological implications, and carry out a synthesis of its constituent elements. This calls for a comparison of the various systems of units of measurements and the choice of the most rational system for the calculation of the results. What should be borne in mind is that the main thing in the solution of a problem is the attentive analysis of its content and the appropriate parts of the solution.

Each stage in the thinking of the student during the solution of a problem requires justifying the choice of the theoretical facts adopted for the solution i.e., concepts, relationships, laws. Only in this way can the solution be properly carried out.

The student will derive satisfaction by solving independently any problem, no matter how simple, provided it is of topical interest. He thus gains confidence in his mental ability and becomes ready to tackle other, more involved problems. The solution of problems is a valuable method for consolidating the knowledge of factual material and offers the means of applying a general physical theory to particular conditions. This association of theory and practice provides a strong stimulus to the development of the reasoning of the students. Of course, the value of this method should not be over-rated. It cannot replace by any means the visual perception of phenomena and independent work in the laboratory.

The use of analogies is another effective tool for the explanation of various physical questions. The understanding of many physical concepts can be enhanced by means of analogies.

The study of physics offers tremendous possibilities for the development of the reasoning of the students. Some concrete examples of this were given above. In every school subject, the procedures and methods of logical reasoning are applied and given substance by means of the material specific to that subject. By virtue of this fact the reasoning itself acquires some specific characteristics. Thus, for instance, reference is made to "physical" reasoning, which implies the ability to isolate the main point in complicated phenomena and disregard incidental details, and of evaluating various relationships quantitatively. This ability is developed to a certain extent in secondary school, where the rudiments of the physical way of thinking are acquired. Its development must be particularly encouraged in students who are interested in physics and engineering. The interest of these students must find an outlet in reading popular science, performing experiments, designing instruments, building models of machines, etc.

The development of reasoning during the study of the fundamentals of physics lays the foundation for further development of the "physical" way of thinking in higher education.

The development of the logical and dialectical reasoning of the students during the study of physics, as of any school subject, is a very complex problem. Unfortunately, this problem has not been adequately discussed and worked out in the methodology of physics teaching in the Soviet Union. Nor does the present paper claim to cover all the ramifications of the subject. It is only through the combined efforts of many physics teachers and methodologists that a proper solution of the problem can be eventually achieved.

B. M. YAVORSKII

"THE ELECTRICAL PROPERTIES OF THE SOLID STATE" IN SCHOOL PHYSICS

Among the various properties of solids, for which ever increasing applications are found in science and technology, an important place is occupied by the electrical properties of metals, insulators, and semiconductors. Academician A. F. Ioffe has observed that in the last 25 years semiconductors have assumed a prominent role in solid-state physics and become the basis of technical progress in such fields as automation, high-frequency radio engineering, current transformation, refrigeration, and heat engineering. The application of semiconductors in these fields has uncovered new, far-reaching, practical possibilities.

At the same time, the current curriculum of school physics makes no provision for acquainting the students with the properties of semiconductors and their applications.

Modern physics has wrought considerable changes in the familiar, "classical" conceptions of the electrical properties of solids. Important aspects, such as the criteria differentiating metals from insulators, the physical meaning of the temperature dependence of the resistance of metals, and many others, have come under a completely new light.

An inadequate amount of space is devoted in the school physics course to the diverse properties of solids, which is not consonant with their topical importance in science and technology. It may appear that this applies to a lesser extent to their electrical properties. Indeed, in the first stage of physics instruction in the 7th grade elementary notions on electric current in metals are given. These notions are further developed and the laws of electric current are studied in the second teaching stage, in the 10th grade, where also some ideas on dielectrics are presented. Unfortunately, however, notwithstanding the considerable time allocated to the study of the electrical properties of solids (in the 1957 curriculum 30 hours were allocated to the study of all topics connected with this aspect), the students are left with many incorrect ideas, e.g., about the intrinsic difference between the properties of solid insulators and metals. On completing the secondary school, they are convinced that metals and dielectrics differ fundamentally in structure. They maintain that in metals there are free, valence electrons which are the cause of conductivity, while in dielectrics the electrons are bound in the atoms and molecules and thus cannot "stream" in a definite direction, i.e., these electrons have no part in electroconduction. Only very few, particularly inquisitive students become aware of the fact that such a marked difference in the structure of metals and insulators ought to produce sharp differences not only in their electrical, but in their other properties as well. Actually, some metals and dielectrics show many similarities in their mechanical, thermal, and various other properties.

Accordingly, legitimate doubts arise as to the validity of the assertion that metals and insulators owe their different properties to a difference in structure.

As we shall subsequently see, these doubts are by no means groundless. The true nature of the difference in the properties of metals and insulators (dielectrics) resides, not in a fundamental difference in the structure of these solids, but in the different behavior of the electrons in the crystals of metals and insulators. The information imparted to the students on these questions does not conform to the contemporary level of development of solid-state physics.

Admittedly, the lack of correspondence between the information the students get in school and the actual state of contemporary knowledge prevails not only in the domain of the phenomena under consideration. There are many aspects in physics which are not dealt with in school up to the modern interpretation of the phenomena, and that has its own reasons. The main one is the fact that many of the ideas and propositions of modern physics are quite complex and lack intuitive appeal, which puts them beyond the level of scholastic achievement and maturity of the students. In spite of this, the physics course (and, to some lesser extent, the chemistry course) in secondary school differs fundamentally from the other departments of natural science, primarily because the development of this course follows closely the evolution of physics as a science. The secondary course in mathematics, for instance, has changed little in structure and content for many decades, and contains "classical" sections which are more or less "permanently set", scientifically; the content of the modern school physics course, on the other hand, must be necessarily revised periodically.

The tremendous importance attributed to physics in modern natural science, and the intimate interpenetration of physics and technology in our time, impose on the constitution of the school physics course some demands which may be difficult to meet: proceeding on the basis acquired in the first stage of instruction in physics and in the study of related fields (primarily mathematics and chemistry) in the senior classes, the students have to be, on the one hand, brought up to an understanding of modern views and ideas in the more important aspects of the physics course, and of their practical applications; on the other hand, the exposition has to be kept within the grasp of the students. The polytechnical education in secondary school lays specific emphasis on the first point. One of the more important aspects of this kind is no doubt the body of ideas bearing on the electrical properties of the solid state.

The present article discusses modern views on the electrical properties of metals, insulators, and semiconductors, and suggests some methodological considerations concerning the possibility of presenting these ideas in the school physics course. The author is quite aware of the fact that, given the present time budget allocated to the second stage of the physics course, any attempt to expand the conceptual framework of the course and of introducing the problems of modern physics will encounter legitimate objections and come up against the question, at the expense of what is this to be done? However the author proceeds from a profound conviction that the situation of the physics course will be modified as a whole. Its study will be allocated the time necessary for the students to master the principles of classical and modern physics, and, armed with a well-defined body of polytechnical knowledge, they would be able to contribute constructively to the various branches of the national economy.

1. The classical theory of electrical conductivity in metals and its difficulties

The classical theory of electrical conductivity in metals is based on the concept of free valence electrons contained in any metal. For instance, when a crystal lattice of metallic sodium is formed, the eleventh valence electron of the sodium is stripped off and becomes a collectively shared electron, not belonging to any particular atom. These free valence electrons move randomly over the metallic lattice sites, among the regularly arranged positive sodium ions. The presence of free electrons in metals is confirmed by experiments involving the effect of the electrons, manifested on stopping a piece of moving metal. The experiments of Tolman and Stuart and of Mandel'shtam and Papaleksi proved that the transient current detectable with a galvanometer when a piece of metal is stopped and due to the motion of charges by inertia relative to the lattice, is produced by electrons. Tenth-grade students are perfectly capable of understanding the explanation of these experiments, with the help of simple diagrams. This fundamental experimental fact, coupled with Riecke's well known experiment showing that the weight of the metal does not change during the conduction of electricity, i.e., that none of the material of the metal is carried off by the current, form the experimental basis on which rests the study of the classical theory of electroconductivity in metals. Let us also mention the important Hall effect, which consists of the appearance of a transverse potential difference in a metal placed in a constant magnetic field. This effect indicates that the electrons are "pushed over" to one side of the metal by the force exerted on them in the magnetic field.

The Hall effect may be used to determine the concentration of free electrons in metals. It is found that univalent metals have one free electron per lattice site. In bivalent metallic copper there are two such electrons, and in trivalent aluminum there are three electrons per lattice site.

The main task of the electron theory of metals was to explain the mechanism of electrical conductivity and to derive theoretically the laws of Ohm and of Joule-Lenz. The basic ideas of the electron theory of conductivity in metals are discussed in the school physics course, in the second stage of study of electricity. It is explained that the effect of the current source is to produce an electric field inside the metal, which causes a "streaming", i.e., a directed movement, of electrons, which is superimposed on their random motion.

In spite of the fact that the orderly motion of electrons occurs at quite low velocities (even at high current densities in the metal) as compared with their velocity of thermal motion, a current is produced in the metal at once*, because the field inside of it is set up virtually instantaneously, and the electrons in any section of the external circuit are set into orderly movement. Resistance is physically interpreted as the collision of electrons with the randomly oscillating lattice sites, and it is pointed out that the energy transferred by the electrons to the lattice causes heat to be released in the conductor.

---
* The questions related to the build-up of a current are partially considered in the study of electromagnetic induction.

The exposition of these questions in accordance with the classical electron theory of metals creates a spurious impression of the success of the theory and of its correspondence with experimental fact. The discussion of the electron theory of conductivity in metals usually ends up with Ohm's law, when actually it should be followed up by an analysis of the degree of correspondence of the results of classical theory with experimental facts. This analysis, which is perfectly within the reach of 10th-graders, should consist of an examination of the difficulties which have cropped up in the classical theory of electrical conductivity and which are subsumed as follows:

1. **The temperature dependence of the resistance of pure metals.** It is well known that in most pure metals the resistance is directly proportional to the temperature, i.e., the specific conductance is inversely proportional to the temperature, thus $\sigma \sim \frac{1}{T}$. This law of dependence of σ on T could not be derived in the classical theory of electrical conductivity, which left the temperature dependence of the resistance of metals unexplained.

2. **The length of the mean free path of electrons in metals.** In the derivation of Ohm's law it is assumed that the conduction (free) electrons move without collisions from one lattice site to another. (The collisions between the electrons may be neglected as they do not affect the movement of the electrons under the effect of the electric field.) Now since the spacing between the lattice points in a metallic crystal is of the order of 10^{-8} cm (the lattice constant), it is natural to think that the mean free path of the electrons in metals should be of the same order of magnitude. It actually turns out that the mean free path is larger by two orders of magnitude, i.e., it is of the order of 10^{-6} cm. The attention of the students must be drawn to this basic fact; an electron may cross without collision hundreds of lattice spacings in the metal, and this does not square in any way with the picture of electrical conductivity evolved earlier.

3. **Comparison of the specific resistance of metals of different valence.** Proceeding from the fact that bivalent metallic copper has two electrons per atom (lattice ion), while in aluminum there are three, it would follow that aluminum has more free electrons than copper so that, other things being equal, the resistivity of aluminum should be lower than that of copper. Experiment shows that the opposite is true. The resistivity of drawn copper is one half that of aluminum.

Within the theory developed before the students the resistivity of polyvalent metals should be lower than that of metals of lower valence. This is contrary to experimental fact.

4. **The greatest difficulty in the classical electron theory of metals was the explanation of the specific heat of metals.** We have left it for the last, as it does not pertain directly to the problems concerning the electrical properties of solids but is of more general significance.

Besides, the exposition of this problem exceeds the scope of the school course, in view of the fact that the presentation of the principles of the kinetic theory does not cover the calculation of the specific heat of gases. Anyway, the point is that the specific heat of metals as measured by experiment is close to 6 cal/g·atom deg, whereas it ought to have been 9 cal/g·atom deg if one proceeds from the idea that, on heating up by one degree, 6 cal/g·atom·deg are "taken up" by the crystal lattice of the metal and 3 cal/g·atom·deg should go to the "electron gas" in the metal. The electrons in the metal do not, in fact, take any part in the specific heat of the metal. To sum up, it may be said that the classical electron theory of conductivity failed to account satisfactorily for all the experimental facts connected with electroconductivity in metals. Its initial assumptions about the electronic nature of the conductivity of metals were correct. However the classical description of the behavior of electrons in solids, and in particular in metals, has proved unsuitable for the explanation of the electrical properties of the solid state. These difficulties should not be glossed over in the school physics course.

2. Some remarks on electrical conductivity in the light of modern physics

The modern theory of electrical conductivity came into being almost directly following the formulation of the principles of wave mechanics. Before proceeding to an appraisal of the theory, let us dwell on the question as to what ideas and factual information from modern physics are needed to grasp the principles of the modern conception of electrical conductivity in the solid state. What we have in mind is this: when reviewing the material of the 10th-grade course, after having studied optics and the elements of atomic physics, it is well to turn back to the problem of electrical conductivity in the solid state. It may be claimed that the review should not include any fresh material, but the objection appears to us not very convincing. After all, the aim of teaching (reviewing included) is to furnish the students with clear and correct ideas on the mechanism involved in the phenomena. We may mention an alternative way of teaching the electricity course in the 10th grade. It is possible, when studying the laws of direct current in metals, to confine oneself first to a phenomenological discussion, without tackling the electronic mechanism of conductivity.

Then, only after having considered the properties of isolated atoms, the students may be taught some of the properties of aggregates of atoms combined together in the solid state. This organization of the material presents several advantages. The main one is that it permits to study not only the electrical but also the optical and other properties of solids at a level consonant with that of modern science and with the importance of the solid state in modern technology.

Leaving open for the time being the question as to which of these two procedures is to be adopted, let us examine the particular concepts of modern physics which have to be imparted to the students for their correct understanding of the electrical properties of solids*, and the possible methodological means of presenting them. These concepts are relatively few.

A principle of prime importance in modern physics is the fact that electrons, atoms, and molecules are endowed with wave characteristics. These characteristics are also exhibited by protons, positrons, and other "elementary" particles. The wave-like character of particles is proved by various experiments, whose nature is well within the grasp of 10th-grade students. Take the diffraction pattern obtained on the scattering of electrons

---

* Let us note that these concepts have a much broader cognitive significance. They are embodied in the foundations of modern physics, with all its vast range of applications.

(as well as atoms and molecules) by a solid surface. After the students have studied the diffraction of light by a slit and a diffraction grating and have been shown the characteristic distribution of the illumination (i. e., the light and dark fringes) on a screen, it should not prove difficult to explain the experiment of Davisson and Germer on the reflection electrons from a nickel surface.

Even closer to the familiar experiments on the diffraction of light are the experiments involving the passage of an electron beam through polycrystalline powders. Such experiments were first performed by Tartakov in Leningrad and by Thomson in England.

After electrons are made to pass through a sheet made out of a polycrystalline powder and hit a screen coated with zinc sulfide, a distinct pattern of flashes is observed where the electrons hit the screen. The pattern shows bright and dark rings, perfectly similar to the diffraction pattern obtained on the passage of X-rays through the sheet. These experimental findings, together with the experiments involving the reflection of molecular beams from surfaces, which proved the diffraction scattering of molecules, firmly established the fact that a wave can in some way be associated with any moving particle.

When considering the photoelectric effect, the students become acquainted with the wave-particle dualism of light. They familiarize themselves with the fact that there are some phenomena, such as interference and diffraction in which light manifests itself as a wave, while there are others, such as the photoelectric effect, in which it displays quantum characterisics. In this case the point must be brought home that either property may be manifested, depending on the wavelength of the light. When we move over the scale of electromagnetic waves, from the longer to the shorter wavelengths, from radio waves through the infrared to visible light, going on the shorter ultraviolet and ending with X-rays and radioactive radiation, a certain regularity can be noted in the way in which the wave and quantum characteristics of light manifest themselves: with the decrease of the wavelength the quantum characteristics are displayed more and more distinctly, while the wave characteristics show up less and less. This is borne out by the fact that the photoelectric effect takes place when metals are irradiated with light of short wavelengths. Also, it proved quite difficult to demonstrate the wave-like nature of X-rays, and this was done only when Laue used for a diffraction grating a crystalline solid with a lattice constant of 10^{-8} cm. Now does gamma and cosmic radiation constitute the limit of short wavelengths ? It is known that Louis de Broglie first drew attention to the fact that there exist even shorter wavelengths, associated with the motion of electrons and other particles. These waves are not electromagnetic, but of a special kind, and the consideration of their probable nature may be omitted in a school course. What is important is to make the students understand the fundamental fact that the motion of electrons cannot be considered in the same way as the translation of bodies in mechanics. Electrons in motion are associated with the propagation of an "electronic wave".

The wave characteristics of electrons and the other particles composing atoms and molecules have provided a fresh interpretation for the behavior of the electron shells in atoms and of atomic nuclei. As will emerge in the following, they have had a tremendous impact on the development of modern ideas on the electrical properties of solids.

Another important proposition of modern physics, which stands in sharp contrast to the usual "classical" conception, is the quantum character, i.e., discreteness, of the variation of some physical quantities describing the behavior of atoms, molecules, and their constituents — electrons and nuclei. In classical physics it was thought that such quantities as the energy, angular momentum, and magnetic moment of an atom may vary arbitrarily and assume all kinds of values, that is to say, they could change in a continuous manner. Now in modern physics, an extensive amount of experimental data has made it necessary to revise these ideas. It turns out that the energy of the electron can assume only definite, so-called quantized values. The electron may stay only at certain energy levels. The basic idea of distinct energy levels for the electrons in atoms is presented in the 10th grade when studying Bohr's postulates. It is explained that the electrons move without radiating along definite orbits in the atom, and that an electron emits of absorbs a quantum of light when it passes from one orbit to another.

Suitably generalized, these considerations are quite sufficient to make the students grasp one of the principle ideas of modern physics: in any system of particles, be it an atom, a molecule, or a solid crystal, there exists a definite series of energetic states (levels) at which the electrons may stay. The school physics course unfortunately dispenses with the experimental facts which directly confirm this idea. A case in point are the experiments of Franck and Hertz involving the passage of an electron beam through mercury vapor in a discharge tube. It may be directly derived from these experiments that a mercury atom can receive from an electron only a well-defined amount of energy, necessary for the atom to make the transition from one energetic state to another. Indirect experimental evidence for the existence of electronic energy levels in the atom is, however, furnished in the school course. When spectra are studied, it is shown that every atom has a distinct individual emission line spectrum. One of the chief characteristics of an atom is its line spectrum. Now how is it produced? Each spectral line is actually produced by a definite transition of the electrons from one level in the atoms to another. This confirms the existence of such energetic levels. Diagrams of atomic energy levels are often drawn in school. The x-axis of the diagram is used to plot the allowed energy states of the atom, with arrows pointing down indicating radiative transitions, and arrows pointing up indicating transitions with the absorption of a light quantum.

The students must certainly be given such diagrams, and it may be a good idea to adduce as an example the actual energy-level diagram of the hydrogen atom.

Lastly, the third proposition of modern physics required for the subsequent presentation of the modern conception of electric conductivity in the solid state is the distribution of the electrons over the energetic states of the atom (or of any other atomic system, i.e., a molecule or a solid crystal). In the present case it proves necessary to bring in material known to the students from the chemistry course. Thus they are familiar with the fact that the atoms of different elements have different chemical properties and that these properties are determined by the number of electrons on the outer electronic shell. It is known, for instance, that these electrons determine the valence of the atom. Now why is the third electron in the lithium atom on the outer shell? Why is it not together with the other two electrons, on the "helium" shell? Why could all the electrons in an atom not be on a single energetic level, the closest possible to the atomic nucleus and thus the stablest?

These questions may be posed to the students, and they are well within their grasp.

The students can quite well understand the elementary explanation as to why there can be no more than two electrons at each energy level.

These facts suffice to make clear to the students all the difficulties involved in the classical electron theory of conductivity in metals. It becomes clear, first of all, that if the electron possesses wave properties the lattice sites do not constitute unsurmountable barriers to it. The electron waves are diffracted around them, and thus the mean free path may be much longer than the lattice spacing.

It is of importance to impart to the students a correct conception of the physical significance of resistance in metals. The idea consists in that the electron waves are scattered by the oscillating lattice ions. When the temperature rises, the scattering becomes stronger and the resistance increases. Modern theory yields a correct quantitative relationship for the temperature dependence of the resistance, in good agreement with experiment. If the students happen to display a particular interest in physics, it may be pointed out to them that, according to modern theory, if the oscillations of the ions were strictly periodic the resistance would be zero.

The teacher should make it a special point to explain that the difference between metals, dielectrics, and semiconductors, does not reside in a difference in internal structure, but rather in the way in which the electronic energy levels are distributed and filled up.

Without going into the band structure of the energy spectrum of crystals, it may be intuitively made clear that a crystal will be a metal if near the occupied energy levels there are free, unoccupied levels to which the electrons can "jump" when acted on by an externally applied field.

Crystals exhibit dielectric properties when their "upper occupied" states are separated from the nearest free states by an interval of several electron-volts and the external field cannot make the electrons take up any energy so as to "stream", i.e., produce a current. It is important to stress the fact that, according to the modern view, the electrons in dielectric crystals are not only not "bound" in the atoms, but are in a sense even too free. They just cannot be set into motion in a definite direction by the application of an external field.

Semiconductors constitute a borderline case in the way their electronic energy levels are distributed and filled up. In semiconductors there is a narrow energy interval between the "upper occupied" states and the unoccupied energetic states. After developing these fairly simple ideas, the teacher may, if possible, proceed to discuss the electrical properties of semiconductors and dwell on their applications.

The presentation of the physics of semiconductors constitutes a topic of individual interest, which is beyond the scope of this article.

E. E. EVENCHIK

EXPERIMENTAL FINDINGS IN TEACHING "HEAT AND WORK" IN THE 9TH GRADE AND "DIRECT CURRENT" IN THE 10TH GRADE

Among the methods of inquiry used for working out the problems of contemporary physics methodology an important role is played by the setting up of pedagogical experiments.

If we derive certain results on the basis of theoretical analysis in some particular field, we cannot be sure of the legitimacy of applying in teaching the method we deem necessary for developing the physical ideas involved, as long as we have not made sure, in teaching practice, that these ideas are correctly assimilated by the students and are within their scope of understanding. Such indeed was the case when a solution was being sought to the problem of teaching thermal phenomena in the 9th grade.

Professor K. A. Putilov's investigations in thermodynamics drew in their time the attention of physicists to the fact that the treatment of thermal phenomena in the university and high-school textbooks needs radical revision. After a series of discussions on the subject in the scientific literature, a well-defined viewpoint emerged.

The reorganization of the heat course bore, obviously, first of all on the secondary school. The university physics courses which have appeared in the last decade reflect to a certain extent the modern views on thermodynamics.

The views conforming to modern physical theories then started gaining currency in the educational and methodological literature for secondary schools. In that respect much is due to D. I. Sakharov, who, in his textbook for pedagogical colleges and as a co-author in the collective textbook edited by G. S. Landsberg, first succeeded in furnishing a practicable solution to the problem.

The theoretical investigation of the problem showed that the reorganization of teaching calls for: (1) introducing into the secondary-school physics course the concept of internal energy; (2) drawing a distinction between the concepts of heat and amount of heat; (3) treating heat as one of the forms of energy transfer and the amount of heat as a quantity measured by the amount of internal energy transferred from one body to another during heat exchange, i. e., without any work being performed.

On working out the method of developing these basic ideas in the heat course, we derived the following hypothesis.

It is necessary first to introduce the concept of the internal energy as the energy depending on the internal state of the body (i. e., the temperature, pressure, state of aggregation, etc.) and only then explain that the change of internal energy is defined by the change in the total, kinetic and potential, energy of the molecules.

This method is dictated by the following considerations: the students, having gone over the mechanics course and learned the concept of mechanical energy, associate with the latter a quantity whose changes are defined by the work performed.

Introducing from the start the concept of internal energy as the energy of interaction of the molecules appears to us unwarranted, since the students have to be shown beforehand that changes in the energy of the body are associated with the changes in the motion and interaction of the molecules. Presenting this without preparation would mean that one proceeds formally, without concerning oneself with the development of the line of reasoning of the subject.

Besides, as is known, the concept of internal energy comprises: (a) the kinetic energy of translational and rotational motion of the molecules; (b) the potential energy of interaction of the molecules caused by the forces of interaction; (c) the energy of oscillatory motion of the atoms; (d) the energy of the electron shells of the atoms; (e) the intranuclear energy; (f) the energy of electromagnetic radiation which fills to a certain density the space occupied by the body. Therefore no error is committed only in the case when not the whole internal energy but only its changes are associated with the changes in the potential and kinetic energy of the molecules.

The experimental teaching of this section was conducted at the two 9th grades (9a and 9b) of Moscow school No. 204, in 1955-56. The teaching was done by the school teacher Z. M. Grudskaya. The presentation of the material by the teacher and particularly the answers of the students were carefully recorded.

In order to prove that changes in the internal state of the body bring about changes in the internal energy, which depends on it, demonstrative experiments were performed involving the adiabatic expansion of air (Figure 1) and with a tube containing ether, from which on heating by friction the cork is thrown off (Figure 2).

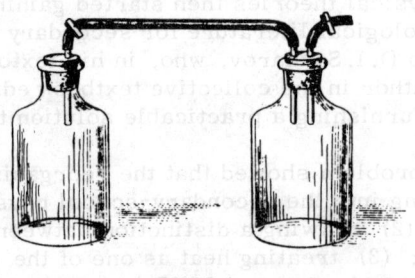

FIGURE 1　　　　　　　　　　FIGURE 2

Interrogation of the students on the demonstrated experiments showed that they had properly understood this basic proposition.

At the following lesson the students were required to describe the experiments they had been shown, which they correctly explained. They referred to the fact that the appearance of mist in the first flask attests to a lowering of the temperature in it, which produces condensation of the vapor inside. The mechanical energy of the system does not change in that case, only a change in its internal energy takes place (the temperature drops). At the same time work is done on the expansion of the air when the clamp shutting the first flask from the second (out of which the air is exhausted by pumping) is released. This means that the work could have been performed only at the expense of the energy determined by the internal state of the body.

It was next necessary to establish the connection between the changes in the state of bodies and the changes in the motion and interaction of the molecules. This required introducing information known to the students back from the 7th-grade course and from the preceding topic, the principles of molecular kinetics.

It is interesting to note that it was possible to clarify the above problems by a heuristic discussion. The knowledge of the students proved sufficient to derive, under the guidance of the teacher, the necessary conclusions to the effect that a temperature rise in the body, which brings about an increase in the internal energy, consists of an increase of the kinetic energy of the molecules, and that a change in the state of aggregation is associated with a change in the distribution of the molecules, and consequently with a change in their reciprocal potential energy.

The concept of heat was explained as one of the forms of energy transfer by comparing the two processes by which the energy can be changed — by work and by heat exchange (i.e., mechanical and thermal). This also has the advantage of extending the concept of work.

The formation of the concept of work in the 8th grade goes through two stages. It is first identified by an unmistakable sign, viz., a force acting on a body over a given path indicates that work is performed ($Fs \neq 0$). Further on, when the concept of energy and the law of conservation and transformation of energy in mechanical processes are introduced, the concept of work is expanded; work is then associated with macroscopic changes in the energy, i.e., with the processes in which macroscopic displacements of the body take place under the effect of impressed forces. This body of concepts is amplified when heat is presented as another way in which energy changes may be effected.

The assimilation of the discussed ideas by the students was checked by a carefully recorded oral examination in class and by having them write a test paper which included questions pertaining to the concepts set forth in the topic "Heat and Work".

Further a record was drawn up of the students' answers on the topics "Changes in the State of Aggregation" and "Work of Steam and Gas", in which these concepts are employed again and help learning the new material.

We give below some student answers on the subject.

Q. Steam at high temperature is kept in a closed cylinder, under a motionless piston. In what ways can the internal energy of the steam be reduced?

A. (student M.) If the piston starts moving, the steam performs work. It will then cool down, just as we saw in the experiment in class, and its internal energy will decrease.

Q. Can the internal energy be reduced without any work being performed?

A. (same student) It can. If the cylinder with the steam is cooled, for instance by pouring cold water on it, the steam will condense and its internal energy will decrease.

A. (student K.) It may be bent and unbent, and it will warm up. Or it could be pounded with a hammer. It could be also placed in hot water, and then its temperature would rise.

Similar answers were given by many students to the question as to how, for instance, the internal energy of a piece of metal (a lead plate) could be increased.

When a cool body is brought into contact with a warmer one, the following takes place: the molecules of the warmer body collide with those of the cooler one and transmit to them part of their energy. Thus the energy of the warm body decreases while that of the cool one increases.

We then say that some amount of heat has been transferred from the warm body to the cool one.

As these answers have shown, the students understand that two kinds of processes may be involved in energy transfer — work and heat exchange.

The comparison between the different forms of energy changes proved quite fruitful, as it made the students appreciate the fact that the magnitude of the work performed measures the change in internal energy produced by the work. They understood, similarly, that the amount of heat measures the change in internal energy occurring in heat transfer (i.e., without work being done).

All students correctly answered the question as to what was meant by the statement that "the body received a certain amount of heat". A summary historical exposition was included in the lessons to explain the origin of the "caloric" terminology.

In their answers the students correctly pointed out that imparting to the body a certain amount of heat means that the internal energy of the body is increased during heat exchange, without work being performed.

The students also properly understood the molecular-theoretical explanation of heat exchange as a microphysical form of energy change (not using that term, of course.)

We give the answer of student K.:

Q. We examined the isobaric expansion of steam and ascertained how this can be achieved. Now you try to tell me on what the amount of heat imparted to the steam during this process is spent.

A. (boy student G.) In the isobaric process the temperature of the steam rises, and its internal energy therefore increases. What's more, the steam performs work by expanding.

Accordingly, the amount of heat received in heat exchange goes to increase the internal energy of the steam, and causes it to perform work by expansion.

The experimental teaching showed that the students, having assimilated the concept of internal energy, were in a better position to graps the idea of fusion and evaporation, and of the work of steam or gas.

The students readily analyzed the changes occurring in the state of aggregation, i.e., fusion and evaporation, and explained that any increase of the internal energy produced in these processes at a constant temperature amounts to an increase of the reciprocal potential energy of the molecules, brought about by a change in their relative positions.

The old way of treating heat as the energy determined only by the temperature and equal to the total kinetic energy of the molecules caused some difficulties when trying to explain the changes in the state of aggregation, for, since the kinetic energy of the molecules does not change during fusion or evaporation, it was not quite clear what happened to the heat imparted in the process. The usual argument, that the energy is expended on the work of disrupting the crystal lattice, does not improve matters because it gives rise to further confusion: since the work performed does not produce an increase in the kinetic energy, the students, not being acquainted with the potential component, find it impossible to explain the conservation of energy. They would thus take the expression "the energy is expended on performing work" as meaning that the energy ceases to exist as such. By the introduction of the concept of internal energy this misconception is effectively dispelled. It was particularly gratifying to hear the students' answers on the topic "Work of Steam and Gas".

In the analysis of isobaric, isothermal, and adiabatic processes, the students independently drew the correct inferences and answered the teacher's questions on the way the law of conservation of energy applied to the changes occurring in the state of a gas, on the first lesson they learned on the subject.

Here are some of the answers given by students to the questions the teacher posed the class during the presentation of the new material.

Q. We examined the isobaric expansion of steam and ascertained how this can be achieved. Now you try to tell me on what is spent the amount of heat imparted to the steam during this process.

A. (boy student G.) In the isobaric process the temperature of the steam rises, and its internal energy therefore increases. What's more, the steam performs work by expanding.

Accordingly, the amount of heat received in heat exchange goes to increase the internal energy of the steam, and causes it to perform work by expansion.

Similar answers were obtained when isothermal and adiabatic processes were analyzed in a like manner.

A. (girl student G.) During an isothermal process the temperature doesn't change, so the internal energy doesn't change, either. All the amount of heat imparted to the steam goes therefore into the work performed in expansion.

The examining board for the matriculation certificate of the class of 1957 noted the ease and assurance with which the students answered the questions on this subject.

In the 1956 - 57 school year the experimental teaching of the topic "Direct electric current" was undertaken under the instruction of the teacher Z. M. Grudskaya, with the same students graduating into the 10th grade, and later, in the 1957 - 58 school year, with the new 10th-grade classes.

This topic does not have to be put to a thoroughgoing experimental test. Many aspects have been adequately worked out and discussed in the methodological literature.

Let us indicate the points which definitely call for new methodological solutions:

1. In recent years it has become common practice to mention in the secondary-school physics educational and methodological literature the necessary presence of extraneous forces in any d. c. circuit and to introduce the emf of the current source as related to these forces. However, optimal methodological solutions are still being sought.

2. Some difficulties are encountered in presenting Ohm's law for a complete electrical circuit, and this point has been often mentioned in the methodological literature. This question involves several methodological problems, whose solution has to be experimentally tried out as well.

3. The rapid expansion of electronics and its industrial applications calls for particular care in developing the students' notions in electronics. One of the major problems is the conceptualization of the electron, and the explanation of the experimental foundations of electron theory together with their subsequent application in the study of the main lines of the subject, including the interpretation of Ohm's law. For the solution of this problem we also resorted to pedagogical experiment.

4. The problem of developing the concepts of electronics is closely related to that of presenting Faraday's laws of electrolysis and of determining the charge of a univalent ion.

The exposition of these questions constitutes an important stage in the formation of the students' ideas on the electron.

At the same time, the contention has been repeatedly voiced recently among some of the physics methodologists and teachers that these questions are too hard for 10th-grade students, that the students cannot grasp Faraday's second law, and that the latter should therefore be excluded from the physics curriculum, as should be also the determination of the charge of univalent ions.

The present writer's pedagogical experience, of many years' standing, has shown that the difficulties the students may have in assimilating Faraday's second law are easily eliminated by adopting the appropriate method of exposition. The didactic importance of the subject appears to us incontestable. In order to prove our assertion conclusively, a teaching experiment was conducted at the school No. 204 for two years, with the help of the teacher Z. M. Grudskaya.

It is beyond the purview of the present paper to discuss the full course of the experimental work and of the findings on all the points specified above. Moreover, the first three of these have been tested out during one school year and will be tried out again next year. We shall, therefore, pause to consider only the last point.

The analysis of personal teaching experience has led us to the conclusion that the difficulties involved in the presentation of Faraday's second law are due to the following causes.

In the exposition of Faraday's laws use is usually made of the chemical equivalent, with which the students are not acquainted in the chemistry course. It is thus up to the physics teacher to explain this concept when presenting Faraday's second law. Actually the introduction and definition of the chemical equivalent can be dispensed with, since it is possible to use instead the concepts of atomic weight and valence, with which the students are familiar. In that case Faraday's second law may be stated as follows: "The electrochemical equivalent of a substance is directly proportional to its atomic weight and inversely proportional to its valence." The students will properly grasp this law only provided the relationships expressed by it are proved and not just dogmatically stated.

If Faraday's second law is presented not on the basis of a logical derivation but by formally verifying the fact that the ratio of the electrochemical to the chemical equivalent (the latter being introduced at that point in the lesson and merely formally learned) is a constant for all substances, the presentation fails to make the point. The students then memorize the law, but do not realize its meaning. As a result they also have trouble understanding the method of determination of the charge of univalent ions, which is based on Faraday's laws.

The logical derivation of Faraday's second law is actually quite simple; the students easily understand it, and, as our experience showed, find interest in it.

For the derivation of the dependence of the electrochemical equivalent of substances on their atomic weight and valence, two examples were considered in class: (1) the passage of current through two voltameters connected in series, whose electrolytes contained univalent ions, e.g., solutions of HCl and NaCl, and (2) the passage of current through two voltameters connected in series, containing electrolytes with ions of different valences (for instance, one with an H_2SO_4 solution and the other with a $CuSO_4$ solution). The analysis of the processes taking place in the voltameters yields the result that in the first case the electrochemical equivalents of the substances are directly proportional to their atomic weights, and in the second that the electrochemical equivalents are inversely proportional to the valences of the substances.

We give some excerpts from the answers of students on the subject (taken from test papers written on this topic).

Concerning the derivation of the dependence of the electrochemical equivalent of substances on the atomic weight, the student L. (10th grade) writes: "When current is passed through two tanks, one of which contains an HNO_3 solution and the other $AgNO_3$, then, if the tanks are connected in series, the current passing through either of them will be the same. The nitric acid and the silver nitrate dissociate:

$$HNO_3 \quad H + NO_3;$$
$$AgNO_3 \quad Ag + NO_3.$$

When current is passed through, the ions start moving in an orderly way: the hydrogen and silver ions move towards the cathode, and the acid radical NO_3 ions move towards the anode. The same number of NO_3 ions will be evolved at the anode in each tank, and therefore the same number of hydrogen and silver ions will also be evolved at the cathode in each of the tanks. But the atomic weight of silver is larger than that of hydrogen, and as a result the amount of silver collected at the cathode will be 108 times the amount of the evolved hydrogen".

This answer clearly shows that the student has understood the principle involved. She expressed herself somewhat inaccurately, though, (saying "when current is passed through the electrolyte the ions start moving in an orderly way", instead of "when an electric field is set up in the electrolyte the ions start moving in an orderly way"), and did not pursue the argument to its conclusion; having shown that the amounts* of the substances evolved at the electrodes are directly proportional to their atomic weights, the student omitted associating these amounts with the electrochemical equivalents of the substances.

In some of the test papers of other students, written on the same theme, this is shown. Thus in the papers of the students Kh., M., and others, a drawing is given (Figure 3) and it is pointed out that when one coulomb of electricity passes through the electrolytes the amounts of the evolved substances are equal to their electrochemical equivalents. Use is made of the notation

$$\frac{m_1}{m_2} = \frac{A_1}{A_2} \text{ and } \frac{\kappa_1}{\kappa_2} = \frac{A_1}{A_2}.$$

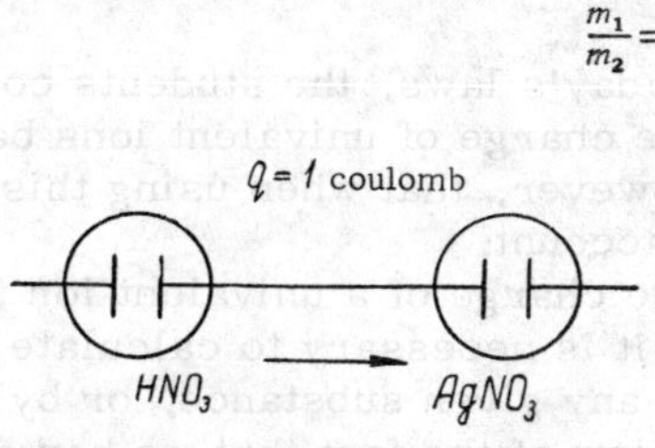

FIGURE 3.

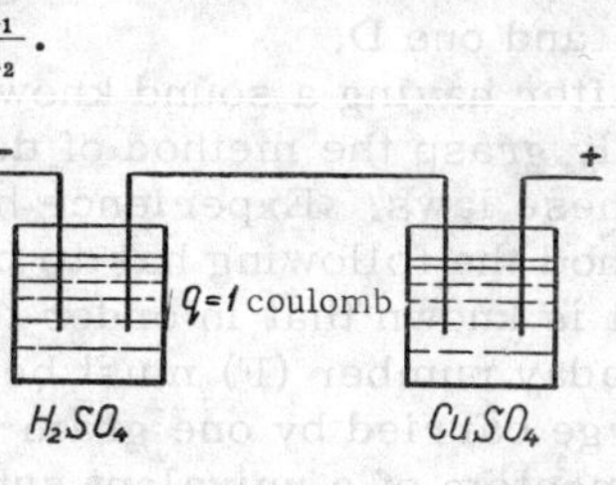

FIGURE 4.

* [This is rather loosely put; the a m o u n t of evolved substance is clearly meant to denote the mass — or the weight — thereof.]

The students also properly dealt with the second, slightly more difficult part, of proving that the electrochemical equivalent of the substance is inversely proportional to its valence.

We give below excerpts from the test papers of the student N., of the same class.

The student made a drawing (Figure 4), and writes:

"The same number of SO_4 ions will collect at the anodes of both tanks, and since the ions are the same their mass in each tank will be the same.*

"The hydrogen and the copper ions will be deposited on the cathode. In the first case, two hydrogen ions are bound with each SO_4 ion, and in the second, one copper ion is bound with each SO_4 ion. Now since the same number of SO_4 ions is evolved in both cases, the hydrogen ions evolved are twice as many as the copper ions."

The student then correctly concludes that the mass of the evolved substance will be smaller, the larger the valence; and since the mass is moreover directly proportional to the atomic weight, the student derives the final result, that

$$m_1:m_2 = \frac{A_1}{n_1} : \frac{A_2}{n_2}, \quad \text{or} \quad \kappa_1:\kappa_2 = \frac{A_1}{n_1} : \frac{A_2}{n_2}.$$

The student L. writes in his paper:

"... twice as many ions of hydrogen are evolved as of copper. The acid radical SO_4 is evolved in equal amounts in both of the tanks.

"The valence of hydrogen is 1 and that of copper is 2.

"We know that when substances of equal valence are evolved,

$$\frac{m_1}{m_2} = \frac{A_1}{A_2},$$

where A is the atomic weight of the substance.

"However, in substances of lesser valence more ions are evolved.

"We can therefore write that

$$\frac{m_{Cu}}{m_H} = \frac{A_{Cu}}{n_1} : \frac{A_H}{n_2},$$

where n_1 and n_2 are the valences of copper and hydrogen.

"As a result, when one Coulomb of electricity passes through the electrolyte, the mass of any of the substances evolved is directly proportional to its atomic weight and inversely proportional to its valence."

Notwithstanding some deficiencies in the manner of presentation, it may be seen from the papers quoted that the students understood the principle involved in Faraday's second law.

The final average grading marks on the test papers in the two experimental classes were as follows:**

10a, 26 students. Among these there were four A's, fourteen B's, six C's, and two D's.

10b, 25 students. Among these there were two A's, thirteen B's, nine C's, and one D.

After having a sound knowledge of Faraday's laws, the students could easily grasp the method of determining the charge of univalent ions based on these laws. Experience has shown, however, that when using this method the following has to be taken into account:

It is known that in order to calculate the charge of a univalent ion the Faraday number (F) must be given, i.e., it is necessary to calculate the charge carried by one gram-equivalent of any given substance, or by one gram-atom of a univalent substance. In view of the fact that we had decided

---

* Secondary reactions in electrolysis were not considered in the physics classes.
** [The grades quoted here have been converted into their equivalents in the American grading system.]

not to make use of the concepts of chemical equivalents and gram-equivalent, to which the students are not introduced in the chemistry course, we resorted to the concept of the gram-atom, which is studied in the chemistry course.

Observations show that the students fail to see the point of inquiring about the amount of electricity transferred through an electrolyte when one gram-atom (or one gram-equivalent) is evolved at the electrodes, if, as is common in teaching practice, the reason for doing so is not explained first. They tend to accept the fact formally and do not perceive why it is important to determine the charge carried by one gram-atom, rather than measure the mass in grams. It is thus first of all necessary to lay the foundation for the choice of this unit of mass.

The students have to be shown that when the amount of electricity carried across an electrolyte by one gram of substance is calculated for different substances by Faraday's first law, a different result is obtained in each case. Their attention is then called to the fact that one gram of different substances contains a different number of atoms in it. Now the gram is a convenient and uniform unit of mass for all substances. However it is possible to choose a unit of mass which, though varying in magnitude for different substances, should contain a constant number of atoms. Such a unit of mass is the gram-atom.

The students know from the chemistry course that one gram-atom of any given univalent substance contains the same number of atoms. Thus, for instance, one gram-atom of copper, equal to 63.57 g, and one gram-atom of chlorine, equal to 35.46 g, are masses differing in magnitude but containing the same number of atoms, equal to the Avogadro number. Accordingly, by determining the charge carried across an electrolyte by one gram-atom of any univalent substance, we actually determine the charge carried by an equal number of ions of different substances.

This makes clear to the students why it is important to be able to calculate the Faraday number (F).

At this point the calculation of this charge and of the charge carried by one univalent atom no longer presents any problem for the students.

The success of the experimental work conducted in school depends in a large measure on the teacher trying out the teaching methods. The unflagging interest and pedagogical tact displayed by the teacher of school No. 204, Miss Z. M. Grudskaya, enabled us to secure results which can be relied upon for working out a method of exposition of the foregoing subjects in the secondary-school physics course.

N. M. SHAKHMAEV

SOME METHODOLOGICAL ASPECTS OF TEACHING ELECTRO-MAGNETIC FIELD PHENOMENA IN SECONDARY SCHOOL

Electromagnetic field phenomena constitute one of the major branches of modern physics. The concept of the "field" itself is a fundamental one, being important not only from the scientific standpoint but also bearing on any outlook on nature. In addition, electromagnetic field phenomena constitute the basis of modern electrical and radio engineering; thus the principles governing these phenomena impinge not only on modern physics but on modern technology as well /11, p. 8/.

However, the study of the electromagnetic field in secondary school at present is not given proper attention. An extensive school survey /5, 6/ has shown that most of the students completing their secondary studies do not possess correct notions on the electromagnetic field and the magnitudes defining it. This was also the conclusion of the representatives giving the university entrance examinations /18, p. 45/. One may find in the educational and methodological literature inaccurate and sometimes basically wrong definitions of the basic concepts and magnitudes describing the electromagnetic field /4, pp. 3,7,73/. It is dogmatically stated in secondary-school textbooks that the electric field is a form of matter, and as concerns the magnetic and electromagnetic fields not even that much is mentioned. Neither is a proper methodological solution provided to the problems relating to the interaction of fields and matter, although this aspect is of prime importance in developing the physical conceptions of the students and in their polytechnical training /6, p. 6/.

The study of electromagnetic oscillations and waves confines itself to the presentation of the most elementary notions, representative of the level of physics at the beginning of the century. Superhigh-frequency electromagnetic oscillations and waves are not studied at all, and as a result secondary-school graduates are left without any proper ideas about the physical basis of television, radar, radio relay lines, waveguides, coaxial cables, modern VHF oscillators, the radiation of electromagnetic waves, etc.

Neither have the proper techniques been worked out for demonstration and laboratory experiments, so necessary for a sound understanding of the properties of the electromagnetic field. Out of a misplaced emphasis on visualization, classroom experiments stand in jarring discord with present-day technology. The assortment of Lodge jars, Ruhmkorff coils, induction machines, Tesla transformers, spark generators, Bronley coherers, and other similar devices making up the "modern" secondary-school laboratory for the study of electromagnetism, is hopelessly out of date.

No solution is provided to a number of purely didactic problems relating to the study of the electromagnetic field. For instance, methodologists are not of a single mind as to whether some elementary notions on the electric field should be introduced in the first stage of physics instruction or not. Also, the way in which the first and second teaching stages are to be interrelated has not been settled yet.

All this indicates that there are many aspects in the method of teaching the electromagnetic field in secondary school which are insufficiently clarified and require further elaboration. With this in view, the author conducted in 1951—1958 a study designed, first, to collect the necessary data for working out an improved methodological procedure, and second, to put the methods developed to an experimental test in school.

In what follows we shall present some of the methodological ideas evolved and describe the new apparatus built in the course of the experimental work.

1. PHASES OF DEVELOPMENT IN THE FORMATION OF THE CONCEPT OF THE ELECTROMAGNETIC FIELD

1st teaching stage

A correct materialistic understanding of electromagnetic field phenomena does not come of itself. A precisely worked out methodological procedure is required, based on a system of suitably selected and carefully interpreted experiments. The teacher must bend his efforts towards having his students achieve a materialistic conception of the field. In this respect it should be borne in mind that any given concept does not come into being all at once but takes a certain time to be formed.

Cognition is not a matter of an instantaneous, inert imprint upon one's mind, but rather a never-ending, self-corrective process progressing from a limited body of knowledge towards a fuller one. From the pedagogical standpoint it is thus necessary to allow some time for the conceptualization of the electromagnetic field. It is advisable to impart elementary notions on the field during the first stage of instruction.

One should not worry about the fact that the students' first ideas on the field might be somewhat inaccurate; this is an inevitable phase in any learning process.

From the methodological viewpoint, the value of introducing elementary notions on the field in the first teaching stage resides in the fact that this furnishes a unified approach to the study of the electromagnetic field in both stages of instruction.

Under the currently adopted procedure, the magnetic field is presented in the first teaching stage, while the electric field is not taught at all. No proof can be found in the methodological literature as to the value of such an approach to electrical and magnetic phenomena, which are actually related to each other. In point of fact, such a procedure cannot stand up to methodological criticism, for, by treating electrical phenomena on an action-at-a-distance basis, we fall into an idealistic, metaphysical position.

As an intersting sidelight, we may mention the fact that in Tsinger's textbook /19,p.221/ elementary notions on the electric field are introduced. There are actually no cogent reasons why elementary notions on the electric field should not be presented in the first stage of instruction, especially now that the field has come to play such a prominent role in physics and technology. These notions are, moreover, necessary for a correct understanding of electromagnetic phenomena.

Any knowledge of our environment takes its origin in sensory impression. V. I. Lenin has pointed out that we have no way of becoming aware of any form of matter or motion other than that of the senses /2, pp. 116 —117/.

The difficulty in studying the electric and magnetic fields resides to a large extent in the fact that neither of these fields affects directly any of the human senses; we are therefore able to ascertan their existence only from their manifestations. This emphasizes the importance of experiments exhibiting the ways in which the electric and magnetic fields manifest themselves.

To display electric field phenomena use may be made of experiments involving such effects as the attraction exerted by an electrified body on other bodies, the attraction or repulsion of electrified bodies, the glow of a fluorescent tube moving in an electric field /3,p.148/, etc. The experiments are used in the first teaching stage just to demonstrate how the field manifests itself, of course without going into the mechanism involved in the effects.

These experiments serve the purpose of bringing home to the students the real existence of the electric field. The demonstration of the spectra of electric fields is of especial importance in this respect. In considering the field pattern the students are presented with a geometrical image of the field, which is quite valuable in enabling them to visualize the field. In order to improve the students' initial understanding of the field, they should be shown by experiment that the electric field is propagated along conductors. It is also useful to establish experimentally the screening effect of conductors and to dwell on the technical applications of the effect.

All this not only endows the field with an objective reality, but later on also simplifies the discussion of the electric current. In particular, it makes it easier to present the concept of voltage potential. The method presently used in the introduction of this concept in the first teaching stage stems mainly from the fact that it is dissociated from the concept of the electric field. The proper treatment of the voltage potential provides, in turn, the means of introducing the concept of resistance in a straightforward and consistent manner.

In developing the students' ideas on the electric field it is well to show by experiment the presence on a field around a two-wire line. This method of approach to the study of electric current not only amplifies the students' understanding of the field but also offers a more accurate way of interpreting the current itself.

The study of the magnetic field in the first teaching stage is given at present considerable attention. However the magnetic field is treated as a space, which does not conform to modern physical views. As in the case of the electric field, it is necessary to show in some simple and easily understandable experiments the ways in which the field manifests itself. In these experiments use may be made of such effects as the action of a magnetic field on steel, the interaction of two magnets, the pattern of fields, the "hovering" of one magnet above another, etc. Like in electrostatics, the demonstration of these experiments should serve to prove to the students the real existence of the magnetic field.

When studying the magnetic field of a current and the phenomenon of electromagnetic induction, the students have to be shown that the electric and magnetic fields are connected with each other. This idea is developed in further detail in the discussion of the second teaching stage. In the first stage the students should be clearly made to conceive the electric and magnetic fields as real entities.

2nd teaching stage

The study of the electromagnetic field at present is subsumed under the headings: "Electric charges. The electric field", The magnetic field and electromagnetic induction", and "Electromagnetic oscillations and waves".

This method of presenting a subject involving one of the major concepts in modern physics cannot be deemed satisfactory, for several reasons:

a. The electric and magnetic fields are studied in isolation from each other and their interdependence is left unexplained up to the last chapter.

b. Direct and alternating currents are studied separately from the field.

c. The part played by the field in electromagnetic processes is inadequately clarified.

Methodologically it is advisable to break down the presentation of the electromagnetic field into the following steps:

1. The study of the electrostatic field in the topic "Electric charges. The electric field",

2. The study of the stationary electromagnetic field in the topic "The laws of direct current",

3. The study of the magnetic field of direct current in the topic "The magnetic field",

4. A first introduction to the variable electromagnetic field in the study of "Electromagnetic induction",

5. The further study of the variable electromagnetic field in the topics "Alternating current" and "Electromagnetic oscillations and waves",

6. The study of the interconnection between the field and matter in the chapter on "Optics and atomic structure".

1. The method of studying the electromagnetic field in the second teaching stage can no longer be the same as in the first stage. The main source of knowledge in the first stage was the bare experiment; now in the second teaching stage, aside from the experiment, stress must be placed on its logical treatment, i.e., on the generalization of the experimental results, as only in this way is it possible to proceed from the specific to the generic — from the extrinsic to the intrinsic aspect of the phenomenon. For instance, there is no experiment by which the boundary of the electric field around an electrified body can be precisely determined; now the analysis of the formula for the strength of the field of a point charge, as derived from experiment, shows that the electric field around a point charge extends to infinity, growing gradually weaker.

The energy of the electromagnetic field is an aspect of particular importance in the correct conceptualization of the field. It is well known that the electromagnetic field possesses energy. Now energy is one of the attributes descriptive of the state of matter and thus cannot be divorced from matter. It may be directly deduced hence that the electromagnetic field is itself a particular form of matter.

Incidentally, in the currently adopted curriculum no mention is even made of the energy of the electromagnetic field. In the study of electromagnetic phenomena references are baldly made to the "transfer of energy", the "radiation of energy", etc. This creates a situation where the students do not associate in their minds the electromagnetic energy of the field, with the result that it becomes a thing in itself, some kind of substance which can be transmitted along wires, radiated as waves, be reflected, etc. This is the way in which energy is in fact interpreted in the current curriculum, under which on two occasions (in the 8th and the 10th grades) reference is made to the transmission of energy.

F. Engels in his time warned against this unwarranted split between matter and energy. As he wrote, the use of "the term 'energy' makes it appear as if energy were something extraneous to matter, something injected into it" /1,p.54/. The students should be therefore shown in some simple experiments that the field possesses energy. Such experiments may be set up as follows.

A length of stiff wire, with an electrostatic pendulum hanging on it, is fastened to an insulating stand (Figure 1). When the wire and the pendulum bob are simultaneously electrified, the bob moves away from the wire and rises to a height h above its original position. In so doing it performs work, $A = mgh$, against the force of gravity. Now since work is a measure of the energy converted from one form into another, it follows that the electric field of the charged bodies possesses energy.

Two carefully washed and dried glass rods are set parallel (Figure 2) and fixed horizontally onto insulating stands. A conductor sphere is placed at one end of the rods (on the left in the drawing) and connected to a high-voltage rectifier or to an electrostatic induction machine.

A lightweight sheet-aluminum ball, B, is put on the rods, so that it touches the conductor sphere, A. When sphere and ball are electrified, the ball moves off along the guiding rods to the point C.

The experiment gives rise to the conclusion that the electric field possesses energy, which in the given case is converted by work against the force of friction into the internal energy of the system.

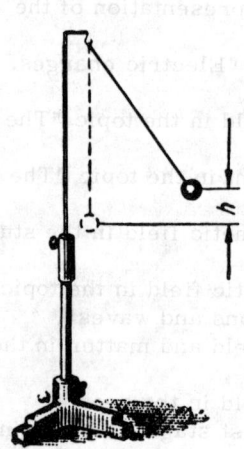

FIGURE 1

The above experiments, simple as they are, provide the means of discussing the energy of the electric field. The principle underlying the experiments is further developed by introducing the concept of potential difference as defining the energy of the electric field.

Later on, when capacitors are studied, it should be shown by experiment that a charged capacitor possesses energy and that this energy is concentrated in the electric field of the capacitor /15, p. 42/.

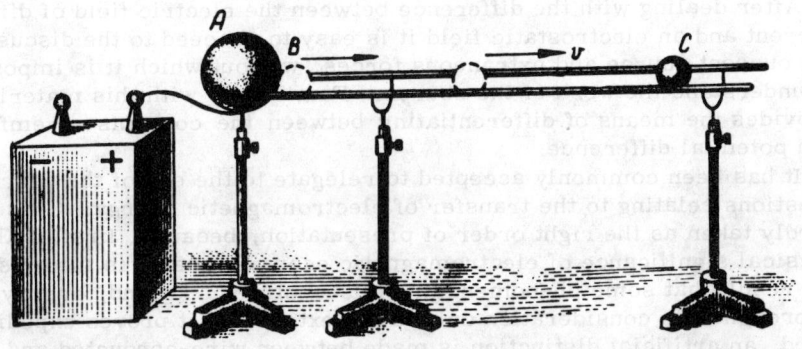

FIGURE 2

In studying the phenomena of electric induction and polarization the students' attention should be called to the way in which the field interacts with matter. In the first teaching stage this aspect is limited to the behavior of conductors and dielectrics in an electric field; the second stage must include a discussion of the electric field produced in these substances.

2. Passing from electrostatics on to the study of direct current proves rather difficult, both for the teacher and the student. This is probably due to the fact that the proper methodological way of dealing with the e. m. f. and the operation of the current source has not been worked out.

The processes taking place in direct-current sources and circuits are considered at present separately from the field. This is hardly an appropriate way of treating the phenomena involved, if only for the reason that any effect cannot be really understood unless its cause is known. A formal treatment of the laws for direct current, isolated from electronic and field conceptions, is not conducive to developing the students' reasoning and broadening their physical outlook but only taxes their memory.

If, on the other hand, the physical phenomena are explained on the basis of the mechanism they involve, the student finds it easier to assimilate the material; this explanation makes the phenomena more readily visualizable and thus relieves the student from having to memorize many formulas and statements of fact and enables him to proceed on his own /8, p. 95/.

The foregoing forcibly proves that it is impossible to content oneself with the prevailing method of presenting the laws for direct current which makes no attempt to take into account the role of the field in direct-current flow. It should be shown by experiment that the electric field is not only responsible for the flow of direct current but that it also pervades the space around the conductors in an electrical circuit. It should be also experimentally shown that in the flow of direct current the electric field is accompanied by a magnetic field /20, p. 86/.

It is very important to show by experiment /9, p. 131/ that the electric field of direct current is different from the electrostatic field which the students had studied earlier. The difference consists in that in order to maintain an electrostatic field there was no need for a continuous consumption (conversion) of energy, while the maintenance of the electric field of direct current requires a continuous consumption of energy.

After dealing with the difference between the electric field of direct current and an electrostatic field it is easy to proceed to the discussion of the current source and extraneous forces, without which it is impossible to understand the work of the source. Familiarity with this material provides the means of differentiating between the concepts of emf, voltage, and potential difference.

It has been commonly accepted to relegate to the end of the course any questions relating to the transfer of electromagnetic energy. This can be hardly taken as the right order of presentation, because, first of all, the physical significance of electromagnetic energy transfer is then lost, with the result that some phenomena are misconstrued; second, energy transfer is brought into consideration only to the extent that it proves expedient; and third, an artificial distinction is made between wire-conducted and wireless transmission of energy.

All this can be avoided if, in the study of direct-current phenomena, the way in which energy is transmitted from source to consumer is examined in broad outline and the energy is associated with the field. This can be done as follows.

A simple electrical circuit, consisting of a source, two wires and a light bulb, is hooked up on a display bench. The students are then asked to answer the following questions: 1) In what direction is the electric energy transmitted? 2) What is the shape of the electric field between the wires? 3) What is the shape of the magnetic field around the wires? Long-standing experience has shown that the students encounter no difficulty in answering these questions. The answers are graphically depicted on the blackboard (Figure 3). It is now required of the students to show the following: 1) The direction of the electric field vector at the point A. 2) The direction of the magnetic line of force at the point A. 3) The direction of transmission of energy at the point A. Their answers are again drawn on the blackboard.

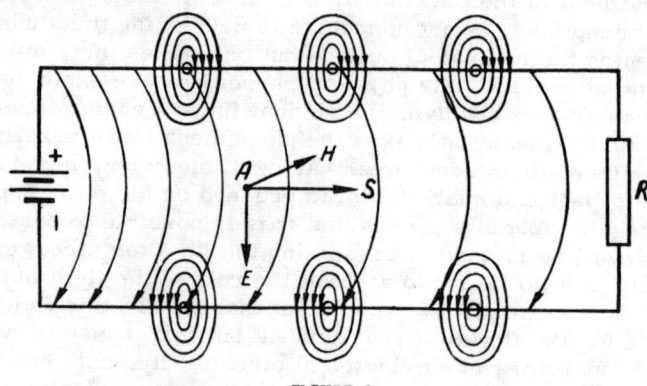

FIGURE 3

The polarity of the current source is then reversed and the above questions have to be answered all over again. It is subsequently deduced that the direction of energy transfer is always perpendicular to the plane containing the electric field vector and the vector denoting the direction of the magnetic lines of force. The students should be told at this point that the vector denoting the direction of energy transfer is known as Poynting's flux vector.*

It must be pointed out to the students that in any instance of electric energy transfer the vector S is always perpendicular to the vectors E and H. Since, however, there are in this case two possible directions, the appropriate one is chosen by means of the screw rule.

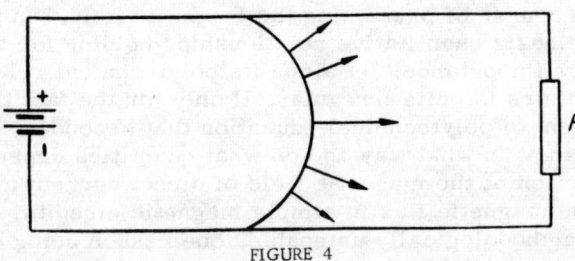

FIGURE 4

Introducing Poynting's vector does not really expand the present curriculum, since the concept is employed anyway in the study of the electromagnetic field /13,p.205/ without actually calling the vector by its name.

After the students have been shown the pattern of the electric field of direct current, their attention is drawn to the way in which the lines of force are curved (Figure 4). Now since Poynting's vector is always perpendicular to the vector E, this means that near the conductor Poynting's vector is directed towards the conductor. Accordingly, part of the energy of the electric field flows into the conductors and there is inside them an electric field carrying this energy. The energy of the electric field within the conductor is transmitted to the electrons in it and by their motion is irreversibly converted into internal energy.

Thus the analysis of the electric field pattern provides the means of demonstrating to the students the fact that the electric energy is carried by the field and that the field is concentrated between the conductors. The energy is accordingly contained in the electromagnetic field between the conductors, not in the conductors themselves. When the circuit is closed the electromagnetic field in a way "slides" along the wires by "leaning" against them. Most of the field energy is propagated in the vicinity of the conductors, where the electric and magnetic fields are the strongest.

Part of the electromagnetic field being transmitted flows into the guiding conductors. The energy of this part of the field is converted into internal energy. The wires of a line are thus not like tubes inside of which energy is transmitted, but act as guides along which the electromagnetic field is propagated.

The laws of Ohm and of Joule-Lenz must be also studied in conjunction with field concepts, as that provides the most simple and also accurate way of analyzing the physical significance of electric resistance and of demonstrating its dual aspect.

* [In the Russian literature usually referred to as the Umov - Poynting vector.]

Consequently, our proposed method for the study of stationary electromagnetic fields under the heading "The laws of direct current" not only improves the students' understanding of the field concept but also furnishes a more accurate way of presenting the laws for direct current.

3. As was the case in the first teaching stage, the magnetic field is studied at present in the second stage only qualitatively. Thus despite the fact that the curriculum prescribes the study of the two characteristics of the field (i.e., the induction and the intensity), no quantitative relationships are derived in the course of presenting the magnetic field. In point of fact, a knowledge of the basic quantitative relationships holding for the magnetic field is of cognitive importance, let alone its polytechnical relevance.

This state of affairs is quite irregular, if only for the fact that it is a specific requirement of polytechnical education that secondary-school graduates should know in what way and on what quantities depend the intensity and induction of the magnetic field of direct current and of a coil, how to compute the magnetic flux in simple magnetic circuits, etc. The situation is also methodologically untenable, one reason being that, in contrast to the magnetic field, the study of the electric field includes the derivation of the relevant quantitative relationships and the solution of a considerable number of problems. This difference in the treatment of either kind of field is unwarranted.

The modification of the method of study of the magnetic field does not call for any expansion of the current curriculum. Let us clarify this by two examples. The present curriculum covers the study of the magnetic field of a coil carrying a current. However, the whole "study" is confined to demonstrating to the students the magnetic field pattern of the coil. It is actually quite easy to show experimentally, by means of a B or H display indicator*, that

$$H \sim \frac{I \cdot n}{l}, \text{ and } B \sim \frac{\mu \cdot I \cdot n}{l}.$$

As a second example, consider the interaction of parallel currents. At present this question is studied only qualitatively. The students are shown an experiment demonstrating the attraction of currents of the same direction and the repulsion of currents of opposite directions. It is actually a simple matter to set up an experiment** from which the interaction law

$$F \sim \frac{\mu \cdot I_1 \cdot I_2 \cdot l}{d}.$$

can be derived.

Proceeding from this law, the magnetic constant and the magnetic permeability are readily defined: this, in turn, simplifies the discussion of the magnetic intensity and induction, whence the basic quantitative relationships for direct current

$$H \sim \frac{I}{d}, \text{ and } B \sim \frac{\mu \cdot I}{d}$$

can be derived. The study of the magnetic field is thus cast into logically consistent system.

We have discussed previously the necessity of presenting the students with the electrical properties of substances. The same agrument applies to their magnetic properties as well.

* The B and H indicators are described further on.
** Described further on.

Magnetic materials are required for the manufacture of various electrical devices, relays, electromagnets, permanent magnets, etc. But despite the fact that magnetic materials play an important part in technology, little attention is payed to them in the school physics course. It should be noted that this field of physics has attained a high degree of development in the last forty or sixty years and has contributed very intersting fresh information on the structure of matter. Nevertheless the properties of magnetic materials are studied in the school physics course on the same level as they were studied ninety—one hundred years ago /6,p. 6/. We can prove this point by comparing the relevant material in a current physics textbook with that in the first Russian textbook in physics.

The textbook of E. Kh. Lenz (1853) contains the following material on the magnetic properties of matter: 1. The action of a magnet on steel. 2) The action of a magnet across nonmagnetic substances. 3) Ampere's theory of magnetism.

In the textbook of K.D. Kraevich (1873) we find the following material: 1) The action of a magnet on steel. 2) The action of a magnet across nonmagnetic substances. 3) The hypothesis of molecular magnets. 4) The behavior of iron in a magnetic field. 5) Paramagnetism and diamagnetism. 6) Magnetic permeability. 7) Ampere's theory of magnetism.

In the standard textbooks of A. V. Peryshkin for the seventh and tenth grades there is the following material: 1) The action of a magnet on steel. 2) The hypothesis of molecular magnets. 3) Ampere's hypothesis.

Comparison shows that the level of study of the magnetic properties of substances in secondary-school is at the same level as it was in 1853 and somewhat lower than it was in 1873. K. D. Kraevich deemed it necessary, back in 1873, to acquaint his students with paramagnetism, diamagnetism, and magnetic permeability, even though magnetic materials had practically no application at the time, while we, in 1958, when it is impossible to imagine modern technology without magnetic materials, do not give such information to the students.

Secondary-school graduates thus do not know the difference between the steel used in the manufacture of permanent magnets and the steel from which transformer cores are made, and do not understand the function of steel in electrical equipment. They do not have a correct conception of ferromagnetism, and notwithstanding the facts they divide all substances into magnetic and nonmagnetic; they fail to differentiate between the intensity and the induction of a magnetic field, they do not possess even elementary notions on the magnetization of a substance, etc.

All this indicates that the study of the magnetic properties of substances must be considerably amplified. This is required in the interests both of physics and of polytechnical education.

The secondary-school curriculum must include as a minimum the following material:

1. The concept of dia- and paramagnetism.
2. The concept of magnetic permeability.
3. The idea of the initial magnetization curve and of the hysteresis loop and the related concepts of magnetically soft and hard materials.
4. The idea of the regions of spontaneous magnetization, without which it is impossible to explain paramagnetism.

Let us note that this material is studied in the secondary-schools in England, France, and Italy much more extensively than is proposed here /10; 16/.

Like in the study of the electrical field, when studying the magnetic field the attention of the students should be drawn to the ways in which the field manifests itself, especially to the processes involving the conversion of of energy. The energy of the magnetic field may be made clear by means of the following experiments.

Two conductors are set in a horizontal position (Figure 5). The conductor A is held fixed, and the conductor B is suspended on a long swing. A pulse of current is let through the conductors, from a bank of storage batteries or of capacitors. Magnetic fields are thus momentarily produced, causing the conductor B to be repelled from the conductor A and to rise to a certain height h, which proves that the magnetic field possesses energy. It must be stressed that the magnetic field derives its energy from the conversion of other kinds of energy into electromagnetic energy (in the current source).

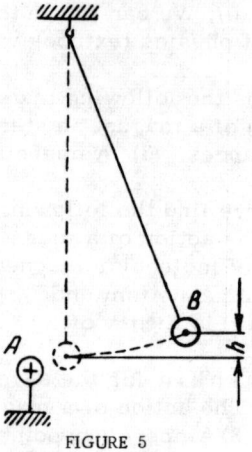

FIGURE 5

The coil of a universal transformer is connected in parallel with a low-voltage bulb to a source of 2—3 V. The voltage of the source is so selected that the bulb filament should not light up. When the circuit is broken the bulb gives a bright flash, and if the inductivity is high enough it may even burn out. In this case the energy of the magnetic field of the coil is converted into internal energy and into energy of radiation. An oscillograph should be used to show how the current varies during the discharge. This will make it possible later on to calculate the energy of the magnetic field of the coil.

4. Under the current method of exposition the electromagnetic field is introduced when studying electromagnetic oscillations and waves. This way of constructing the course produces misconceptions in the students as to the nature of the electromagnetic field, as in this case the student is immediately presented with the study of an alternating electromagnetic field.

By the method proposed above, however, the concept of the electromagnetic field is first introduced in the study of the laws of direct current. When studying these laws, the students learn at the same time about the stationary electromagnetic field, in which the field strengths E and H are independent of the time.

Only after having explained the fundamental properties of the stationary field is it necessary to go on to alternating fields, when studying electromagnetic induction.

Electromagnetic induction can be usually divided into two separate cases, viz., induction in moving conductors and induction in motionless conductors.

Induction in conductors moving in a constant magnetic field is explained as being due to the action of the magnetic field on the moving electrical charges (Loretz force), while induction in conductors remaining motionless in a variable magnetic field is explained as being due to the action of the electrical field induced by the variations in the magnetic field.

Now there is no objective difference between these two forms of induction /14,pp.354—355/, and the students fail to understand why the same phenomenon is explained in two different ways.

The necessity of eliminating this basic difference in the interpretation of two phenomena which are objectively indistinguishable from one another was first pointed out by A. Einstein /17,p.48/. Notwithstanding the fact that this problem has been resolved in the theory of relativity, demonstrating the mutual consistency of the explanations given above is hardly within the reach of school-boys. Electromagnetic induction in conductors must therefore be explained from another standpoint.

In order to develop the students' ideas on the field, in the first teaching stage it is useful to consider induction in conductors moving in a magnetic field as an empirical fact; in the second teaching stage, induction in motionless conductors should be considered from the standpoint of Maxwell's theory, i.e., as being due to the action of the electrical field induced by the variable magnetic field.

This method of approach to the study of electromagnetic induction, even though incomplete, does not imply any inconsistencies.

Furthermore, it should be taken into account that this manner of presentation requires less effort on the part of the student and places less of a burden on his memory.

After having studied electromagnetic induction the students should have a clear idea of the fact that an **electrical field may be induced not only by electrical charges but also by a variable magnetic field.**

5. Further on, when studying alternating current and electromagnetic oscillations, the students will see that also a **magnetic field may be induced not only by moving charges but also by a variable electrical field.**

We may mention at this point that it is not necessary to introduce in secondary-school the term "displacement current", as the word 'current' as applied to the variations in the electrical field produces an incorrect idea in the students' mind, making them identify displacement currents with the more familiar to them conduction currents. It is better to employ everywhere instead of the term "displacement current" the bulkier but more accurate expression "variation of the electrical field." This is chiefly conditioned by the fact that the displacement current and the conduction current are quite distinct physical concepts /14,p.407/.

As our observations have shown, the greatest difficulty in the study of alternating electromagnetic fields crops up in questions involving the radiation of the electromagnetic field (electromagnetic waves). The complications are not caused by the difficulty of the material itself, but rather by the method of exposition. Up to that point the students have been under the impression that the carriers of energy down the conductors were the moving electrons, while in the case of an alternating electromagnetic field it turns out that electromagnetic waves are moving.

This confusion need not arise in the students' mind if the electromagnetic field is consistently kept in view throughout, beginning with a direct-current circuit and ending with a radiating antenna.

It is a good idea to examine with the students the way in which energy is transmitted by an alternating current along a "short" and a "long" line, in the case of a line with a resistive load only.

In the first instance, during the time it takes the electrical field to travel from the beginning to the end of the line the voltage at the beginning of the line remains almost unchanged. At any given instant, therefore, the electrical and magnetic field are the same at the beginning and the end of the line.

When the circuit has no reactance, the voltage and the current coincide in phase over the line, and consequently the variations of the electrical and magnetic fields around the line are also in phase.

In that case the vectors E and H pass through zero and reverse their direction in space at one and the same time. As a result, Poynting's flux vector does not change its direction in space.

Thus, in spite of the fact that the electrical and magnetic fields change their direction in space, the energy keeps flowing in the same direction, just as in the case of direct current.

The only difference is that in the case of direct current the magnitude of Poynting's vector remains constant in time, while in the case of alternating current it periodically varies, attaining a maximum and falling to zero twice within each period.

This means that in direct current the energy flux down the line does not vary with the time, whereas in alternating current it pulsates with time.

It should be repeatedly stressed that, just like in the case of direct current, the alternating electromagnetic field is propagated not in the conductors but along them, with the conductors acting as guide lines.

In the case of a "long" line, the time it takes the electromagnetic field to travel down the conductors is comparable with the period of the electromagnetic oscillations and then either a standing wave (given a reactive load) or a travelling wave (given a resistive load) is set up in the line.

Knowledge of this material permits fairly simply to explain the way in which the electromagnetic field is radiated by a linear half-wave oscillator, proceeding to the latter from a two-conductor line /12,p.80/.

It should be explained to the students that the function of an antenna is to change the configuration of the electric and magnetic fields in such a manner that the electromagnetic field moves away from the conductors into surrounding space.

After the students have understood the way in which the electromagnetic field is radiated, its properties have to be studied by experimental means

The most important aspects, with respect both to physics and to practical applications, are: the reflection, refraction, interference, defraction, and polarization of electromagnetic waves. In order to demonstrate these properties of electromagnetic waves we built a microwave oscillator (its construction is given further on). This oscillator provides the means not only of displaying the effects mentioned above, but also of performing some other interesting experiments (the focusing property of lenses, the passage of waves through a prism, standing waves, an experiment on interference, similar to Fresnel's optical experiment, and some others). The demonstration of these experiments makes it possible not only to study thoroughly the properties of a **free** electromagnetic field, but also to link the electromagnetic field with **optics**, which terminates the school stage in the formation of the concept of the electromagnetic field.

6. In order to ensure that the students have a correct conception of the electromagnetic field, attention should be given in the study of optics to the following properties of light.

1. The finite velocity of propagation of light. The equality between the velocity of propagation of light and the velocity of propagation of the electromagnetic field. The transverse nature of light waves.

2. Light pressure. From the fact that light exerts pressure it necessarily follows that light possesses some momentum.

3. Light energy. From the fact that light carries with it a certain amount of energy and possesses momentum it follows that light possesses some mass.

This information is sufficient to enable the student to draw at least two conclusions: 1) light consists of electromagnetic waves; 2) electromagnetic waves are a form of matter, existing side by side with substance.

In the study of atomic structure the students come across an instance of the conversion of a field into matter, in the electron-positron pair production; the reverse process, of pair "annihilation", exemplifies the conversion of matter into field. This constitutes a proof of the reciprocity and interchangeability of the two forms of matter — substance and field.

This concludes the school stage in the conceptualization of the electromagnetic field.

II. DEMONSTRATION EXPERIMENTS WITH THE ELECTROMAGNETIC FIELD

1. New instruments and set ups

The effective formation of a correct conception of the electromagnetic field is impossible without study by experiment. The fundamental properties of the field have to be experimentally demonstrated. However, the equipment of school laboratories suffers from a certain lack in demonstration devices. Consequently, in the course of work we had to create some new instruments and bring up to date the existing ones.

A special place among these is occupied by the instruments intended for the demonstrative measurement of the quantities E, B and H. As has been shown by a scientific-methodological analysis of the basic demonstrations, these instruments must give not the absolute but only the relative values of the measured quantities, i.e., provide a quantitative standard of comparison of the intensity (or the induction) at different points of the field. These devices are, basically, indicators of the appropriate quantities.

The indicator of electrostatic field intensity (Figure 6) consists of an almost balanced pointer made of Plexiglas, to the one end of which a hollow ball of sheet aluminum is attached. The use of Plexiglas makes it possible for the ball to retain its charge for a considerable length of time when it is electrified.

When the electrified ball is placed in an electrical field it experiences a force proportional to the field intensity. The action of this force makes the pointer rotate about its axis. A counteracting torque is provided by making the bottom part of the pointer slightly heavier than the top. The scale of this instrument, like those of the instruments described further on, is graduated in arbitrary units.

By means of this indicator it is possible to show, with an accuracy quite sufficient for demonstration purposes, that the intensity of the electrical field around a small electrified ball (point charge) is inversely proportional to the square of the distance.

If the indicator is placed between the plates of a flat demonstration capacitor, it is possible to show experimentally that the field intensity is constant on the straight line joining the centers of the capacitor plates.

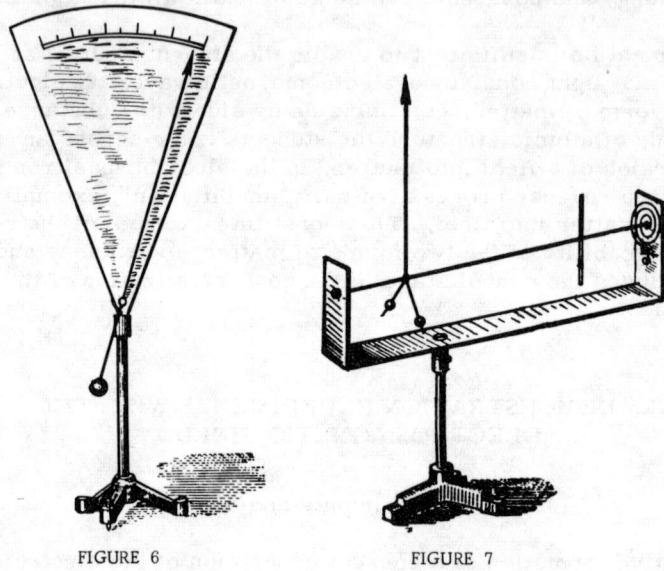

FIGURE 6   FIGURE 7

The indicator of the magnetic field intensity (Figure 7). From the general physics course /14, p. 346/ it is known that the magnetic field intensity may be measured by its effect on a magnetic needle. Thus for the sensitive element in the indicator of the magnetic field intensity use is made of a slim magnetized rod balanced on a horizontal axis. An aluminum pointer is also balanced on the same axis as the magnetic needle. A counteracting torque is produced by a flat coil spring.

In order to make it possible to introduce the indicator inside demonstration coils and in the gap between the poles of magnets, the instrument has a long axis and the pointer is set 100 millimeters away from the magnetic needle so as to give it clearance to rotate. The deflection of the needle is proportional to the magnetic field intensity.

By means of this indicator it is possible to show that the magnetic field intensity depends on the strength of the current, the shape of the conductor, and the distance from the conductor, and that it is independent of the magnetic properties of the medium in which the magnetic field is set up.

The indicator of the magnetic field induction (Figure 8). It is known that the induction of the magnetic field can be determined by the effect of the magnetic field on a current. Thus for the sensitive element in the induction indicator use is made of a small coil fed from a direct-current source. An aluminum pointer is fixed and balanced on a common axis with the armature. The current is fed to the coil through flat spiral springs which serve at the same time to produce a counteracting torque.

Under the effect of the magnetic field to be measured the coil rotates through a certain angle proportional to the induction strength /14, p.346/.

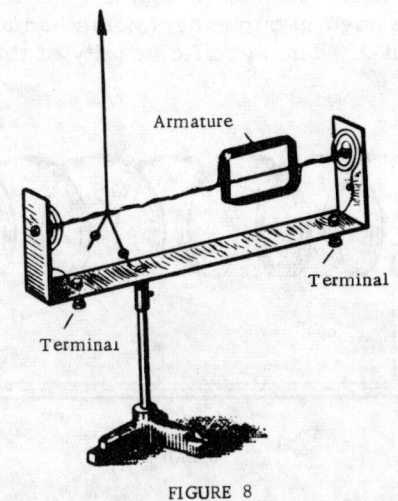

FIGURE 8

The armature is made of such dimensions that the measured field within it may be considered homogeneous. As in the case of the field intensity indicator, the pointer is removed at a distance of 100 mm from the sensing coil.

By means of the indicator of the magnetic field induction it is possible to show that the induction of the magnetic field depends on the current, the shape of the conductor, the distance from the conductor, and on the magnetic properties of the medium in which the magnetic field is set up.

2. Paramagnetic media

In order to demonstrate the dependence of the force of interaction between parallel currents on the magnetic properties of the medium has to be made of a paramagnetic liquid, whose magnetic permeability is 1.5 to 2 times higher than the magnetic permeability of air. Such a liquid does not exist in nature. Even the most paramagnetic liquid — a saturated aqueous solution of ferric chloride ($FeCl_3$) — has a magnetic permeability which is very close to that of air. We prepared a paramagnetic liquid by adopting the procedure used for obtaining magnetodielectrics. A fine ferromagnetic powder was mixed into glycerine. The grain size of the powder must be of the order of a few microns (to prevent it from settling too rapidly). Such powders are available from industry; the most common is magnetite powder. Instead of glycerine use can be made of a lubrication oil of fairly low viscosity, and the manufactured magnetite powder may be replaced by iron oxide or very fine iron filings. In the last case, however, it should be borne in mind that the filings tend to settle rather quickly, and therefore the paramagnetic liquid must be prepared immediately before the demonstration.

The magnetic permeability of the resultant medium depends on the amount of iron powder mixed into it, and is constant in not very strong fields. The medium used in our experiments had a relative magnetic permeability of about 2. The specific gravity of the liquid was approximately 1.5.

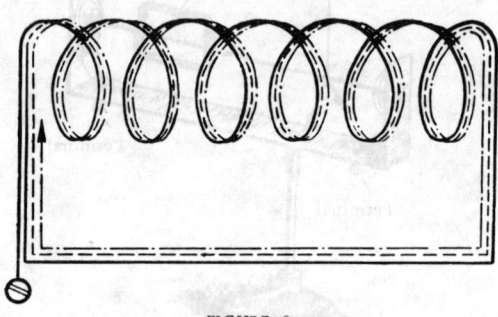

FIGURE 9

Demonstration coils. For the study of the magnetic field of a coil, as well as for some other experiments, two demonstration coils were made. The winding diagram of the coils is shown in Figure 9. Each coil has ten distinctly visible "loops" 100 mm in diameter, and each individual loop is made up of about one hundred turns of 0.5 mm wire. The bottom connecting wires are laid in a groove in the support. The ends of the windings are inserted into universal plugs.

In order to demonstrate the technical application of electromagnets we built demonstration models of an electromagnetic relay, a magnetic starter, and a reverse-current relay.

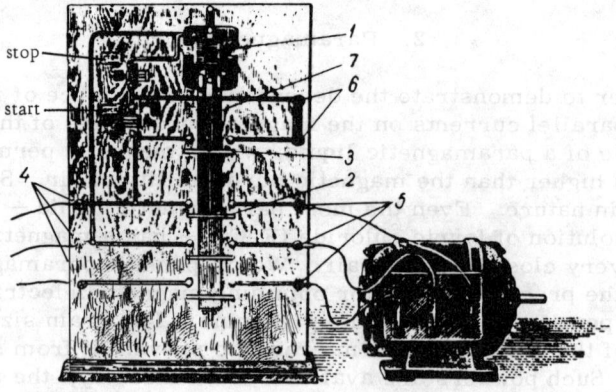

FIGURE 10

The demonstration magnetic starter. The demonstration magnetic starter (Figure 10) is assembled on a vertical insulating panel

350×450 mm in size. The assembly consists of the following components: a small electromagnet 1, a three-pole contact switch 2, a locking contact 3 with compression springs ensuring a tight contact, and "start" and "stop" buttons. The starter is connected to a three-phase circuit by means of the terminals 4, and to the motor by means of the terminals 5. In addition, there are on the panel terminals 6 for connecting the starter with a thermal relay, and when the starter is demonstrated without the relay the terminals are shorted across. In a normal position the "start" button is off and the "stop" button is on.

The wiring is done with heavy insulated wire. The power circuit, the control circuit, and the locking circuit are hooked up with wire of different colors. Figure 10 shows the motor connected into the circuit through the magnetic starter.

When the "start" button is depressed the current flows from the first phase across the "start" and "stop" buttons through the electromagnet winding, and out the second phase. At the same time the electromagnet is activated and moves the draw bar 7 connected to it which closes the power circuit and the control circuit. Releasing the "start" button does not break the power circuit, since the control circuit remains closed through the locking contact 3 which is connected in parallel with the "start" button.

In order to break the power circuit the supply circuit of the electromagnet must be broken, and this is achieved by depressing the "stop" button. When the electromagnet circuit is broken the armature of the electromagnet reverts to its initial lower position under the effect of gravity and of the compression springs. This breaks also the power circuit and the control circuit.

Depressing the "start" button connects the system again. The electromagnet winding is calculated to make the magnetic starter respond to a voltage of 110—220 V.

The magnetic starter may be used for the automatic switch off of the motor when it is overloaded for a long time. Figure 11 shows the motor hooked up to the magnetic starter through a thermo-relay.

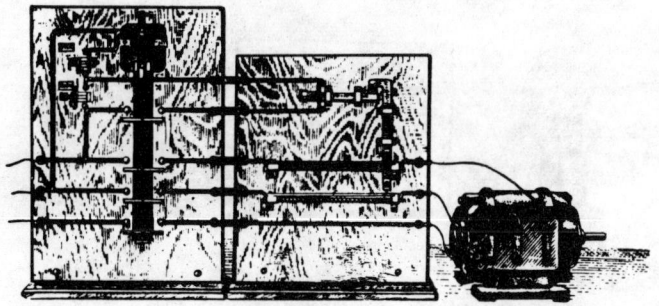

FIGURE 11

The reverse-current relay. The demonstration model of a reverse-current relay (Figure 12a, b, and c) is hooked up on a vertical insulating panel measuring 350×450 mm. It consists of an electromagnet 1 with two windings A and B, a pivoted armature 2 which is pulled apart by the spring 3. The electromagnet windings are made of enameled wire. The first consists of about 1000 turns of 0.25 mm wire, and the second of 20 turns of 1.25 mm wire.

The relay is intended to be used with a school magneto, which is connected to the terminals 4. A storage cell is made to serve at a direct-current source. It is connected to the terminals 5.

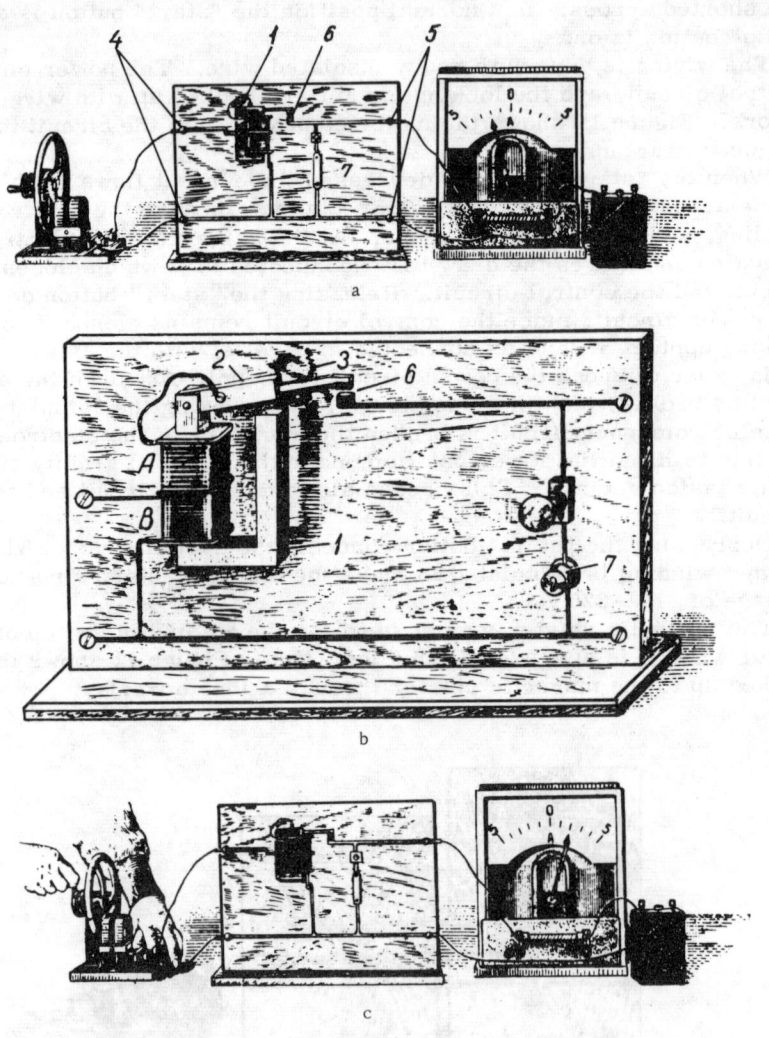

FIGURE 12

When the magneto is not working or is working at low revolutions the light bulb is fed from the storage cell (Figure 12a). When the magneto is turning at normal speed the relay is activated, and the current from the magneto flows across the contact 6 to the lamp and to the storage cell and charges the latter (Figure 12b). In this case the magnetic fields of the coils A and B intensify each other. When the rate of rotation of the magneto is reduced (or when it is stopped) its voltage drops and then the current starts flowing from the storage cell through the bulb and the magneto; in this case the magnetic fields of the coils A and B are opposed and weaken each other. The armature 2 is then pulled away from the electromagnet core by the spring 3, the contact 6 is broken, disconnecting the magneto, and the current from the storage cell feeds only the light bulb. The load may be disconnected by means of the switch 7.

A set of demonstration devices for the study of the properties of electromagnetic waves

A set of the following devices was designed for the study of electromagnetic waves:
1. A microwave generator ($\lambda = 3$ cm).
2. A microwave receiver.
3. A multivibrator for modulation.
4. A repeller-plate current rectifier.
5. Metal plane mirrors (two).
6. A 3 x 30 cm metal plate.
7. A prism out of organic glass, filled with glycerine or water.
8. A grating polarizer.

In order to set up experiments with the above items use is made of the following instruments available in the secondary-school laboratory equipment:
1. A kenotron rectifier.
2. A low-frequency amplifier*.
3. A dynamic speaker on a support*.
4. A demonstration galvanometer.
5. Universal stand supports.
6. A television lens.
7. A concave mirror.
8. An aquarium.

By means of this set the following demonstration experiments can be performed:
1. The radiation of electromagnetic waves by an oscillator.
2. Reflection at the boundary between a dielectric and a conductor.
3. The refraction of waves at the boundary between two dielectrics.
4. The screening effect of a conductor.
5. The laws of reflection.
6. Interference of electromagnetic waves from two mirrors (similarly to Fresnel's optical experiment).
7. Diffraction from one slit.
8. Diffraction from two slits.
9. Standing waves.
10. Diffraction from a rod.
11. The passage of waves through a lens.
12. The passage of waves through a prism.
13. Reflection from a concave mirror.

---

* A radio set fitted with adapter jacks may be used instead.

14. The passage of waves through a plate with parallel faces.
15. Polarization of electromagnetic waves.
16. The principle of radar.
17. Modulation and demodulation.

The demonstration of the experiments given above provides a secure bridge towards the study of optical phenomena.

Description of the basic units of the set

1. The **microwave oscillator** (Figure 13) is built around a reflex klystron of the K-19 type* (any other klystron with similar parameters may be employed). The unit consists of the oscillator proper (the klystron), a resonant waveguide, and a horn antenna. The oscillator is fed from a kenotron rectifier of the kind available in school. The negative voltage for the repeller plate is supplied by a low-power crystal rectifier. The modulating voltage is fed to the repeller plate from a multivibrator. An audio-frequency oscillator can be used instead of the multivibrator.

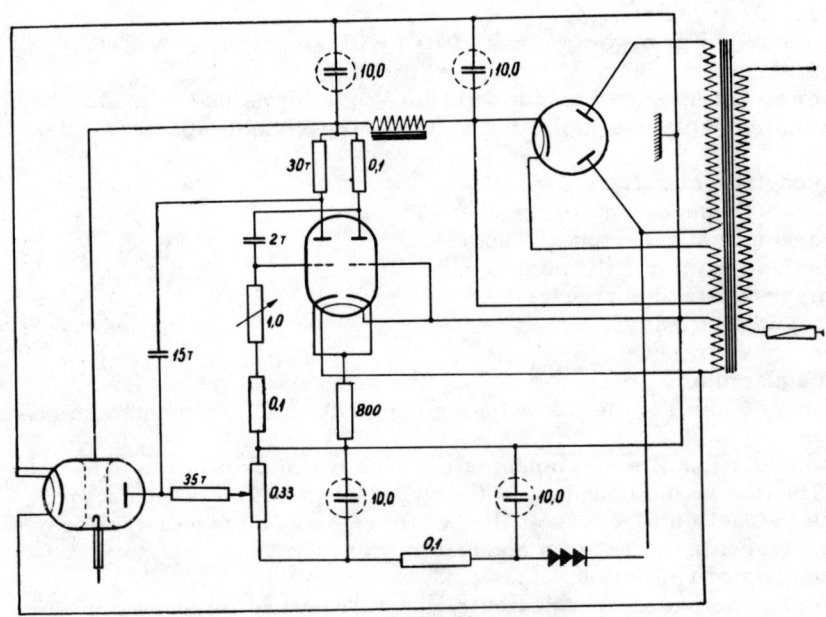

FIGURE 13

2. The **multivibrator** is fed from the main rectifier and is mounted on the same panel as the repeller-plate rectifier. During the demonstrations it is attached to the chassis of the main rectifier by means of the terminals on the latter.

3. The **microwave receiver** is made up of a detector set and a resonant waveguide which is fitted with a horn antenna. As an indicator

---

* [Tube designations are transliterated from the Russian. Their American or British equivalents may be found in handbooks.]

use may be made in the receiver (through an amplifier) either of a dynamic speaker (for modulated signals) or of a demonstration galvanometer. Both the receiver and the oscillator are free to rotate about the longitudinal axis of the waveguide through 360°.

FIGURE 14

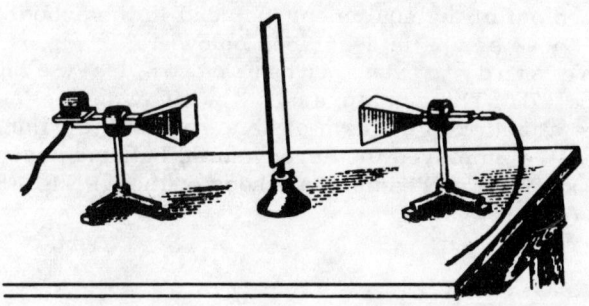

FIGURE 15

FIGURE 16

4. **Individual components:** a) the plane mirror may be made of any metal. The mirror measures 20 x 30 cm, and has at its bottom a metal rod for attaching it to a stand; b) the metal plate measuring 3 x 30 cm is used in the experiments on diffraction, and also has a rod for attaching it to a stand; c) the polarizer is set in a wooden frame which is adapted

to be rotated about its center of symmetry; d) the prism and the lens are made out of organic glass and are filled with water or glycerine. When testing the samples we did not have the possibility to ascertain the optimal dimensions of the prism and the lens. We made use of a television lens, and as a prism we used the corner of an aquarium. Figure 14 shows the arrangement for observing the interference of electromagnetic waves from two mirrors, Figure 15 shows the arrangement for observing defraction from a rod, and Figure 16 shows the arrangement for observing diffraction from a slit.

The layout presented here differs from similar layouts (for instance, that of B. C. Zvorykin) by the small wavelengths involved, which provides the means of demonstrating in a simple and straightforward manner the fundamental properties of waves. This layout makes it possible to set up demonstrations involving the diffraction, interference, and polarization of electromagnetic waves, which are either impossible or difficult to demonstrate on other available layouts. A further distinctive feature of this layout is the use of a klystron, a waveguide, and a horn antenna, and the extensive application of the instruments available in the school physics laboratory.

In addition to the devices described above, we developed some setups which are assembled out of the equipment included in the school physics laboratory. On of these setups is described below.

Setup for the study of the interaction between parallel currents (Figure 17). The armature of a Khazanov device [a form of rectangular coil] is attached to the arm of an aerodynamic balance by means of a holder. We employed the aerodynamic balance designed and constructed at the Institute of Teaching Methods of the APN RSFSR, under the direction of A. A. Pokrovskii.

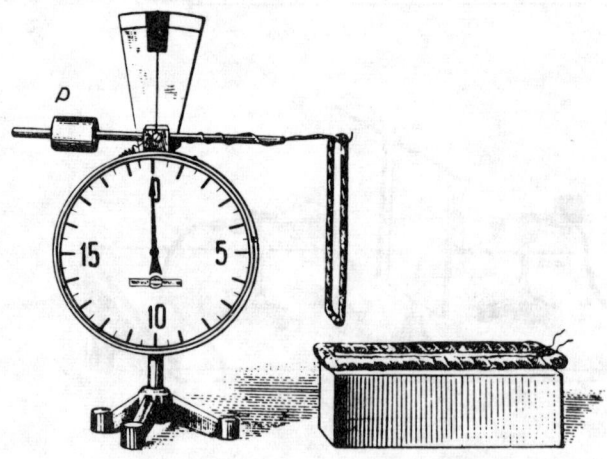

FIGURE 17

The armature is balanced by moving the counterweight P (the pointer of the dynamometer is ajusted to the zero point). A specially prepared frame, consisting of 40 to 50 loops of 0.5 mm wire, is placed on a rest under the balanced armature, at a distance of 0.5 cm from it.

The frames are connected into a circuit so that the conductors should repel each other. Storage batteries may be used as current sources. In order to vary the current in the frames they are connected in series with rheostats having a resistance of 30 to 40 ohms and working on a current of 3 to 4 amp. It is necessary to hook up in the circuit of each frame a demonstration ammeter so as to be able to measure the current.

The two circuits are closed and the current in the two frames is set at 1 amp. The armature is balanced on the dynamometer again and the force of interaction is noted on the scale in arbitrary units. On raising the current in the suspended armature to twice its previous value it is found that the interaction force increases by a factor of 2; when the current is raised to three times its value, the force of interaction also increases by a factor of 3. The experiment thus shows that the force of interaction between parallel current is proportional to the current in either conductor. This result is written on the blackboard as follows:

$$F \sim I_1.$$

By varying in a similar manner the current in the fixed conductor, it is established that

$$F \sim I_2.$$

On increasing the distance between the conductors to twice, three times, and four times its value it is found that the force of interaction is reduced by the corresponding factor, i.e.,

$$F \sim \frac{1}{d}.$$

If the position of the two frames is changed, so that their shorter sides are interacting, it is established that

$$F \sim l.$$

A photographic tray is next placed under the suspended frame; the fixed frame is placed inside and paramagnetic liquid is poured until the two interacting conductors are submerged. The suspended frame is balanced and the same current as before is passed through, whereupon it is ascertained that the force of interaction has increased. It is thus deduced that the force of interaction of parallel currents depends on the magnetic properties of the medium; this may be written in the following form:

$$F \sim \mu.$$

Collecting the observational results, one gets

$$F \sim \frac{\mu \cdot I_1 \cdot I_2 \cdot l}{d}$$

or, passing on to an equality,

$$F = k \frac{\mu \cdot I_1 \cdot I_2 \cdot l}{d}.$$

The resultant relationship makes it possible to introduce in a straight-forward manner the concept of the magnetic constant and the magnetic permeability of the medium, as well to demonstrate the dependence of the induction of the magnetic field on the properties of the medium. Further, it is possible to derive the expression for the induction of the magnetic field of direct current, i.e.,

$$B = k \cdot \frac{\mu I}{d}.$$

Finally, and this is very important, knowledge of the relationship expressing the force of interaction of parallel currents lays the foundation to the definition of the unit of current, the ampere, as is the current procedure in metrology.

The foregoing attests to the importance of the demonstration described above.

It may be pointed out in conclusion that both the setup itself and the experiments conducted with it are simple, possess visual appeal, and are understood by the students.

The methodological ideas propounded above were subjected to repeated tests during 1951—1958 at school No. 215 in Moscow. Throughout the experimental work in the 7th and 10th grades there were no students who failed in physics. The grading marks in these grades were considerably higher than they were in the 6th, 8th, and 9th grades with the same students. This gives grounds to the claim that the method proposed here not only promotes the formation of scientifically correct ideas on the electromagnetic field but also ensures a sound body of knowledge.

BIBLIOGRAPHY

1. Engels, F. Dialectics of Nature [Russian translation.] 1952.

2. Lenin, V.I. Sochineniya (Collected Works), Vol. 14, 4th edition.

3. Bakushinskii, V.N. Analiz fizicheskikh opytov i priborov (Analysis of Physical Experiments and Instruments). — In: Izvestiya APN RSFSR, Vol. 56, Moskva. 1954.

4. Bron, O.B. Pole kak vid materii (The Field as a Form of Matter). — In: Elektrichestvo, No. 7. 1954.

5. Galanin, D.D. and V.F. Yus'kovich. Ob uluchshenii prepodavaniya fiziki v srednei shkole (Improving the Teaching of Physics in Secondary School). — Uchpedgiz. 1951.

6. Dorfman, Ya.G. Sovremennye predstavleniya o magnitnykh svoistavkh veshchestva (Modern Concepts on the Magnetic Properties of Matter). — In: Fizika v shkole, No. 2. 1949.

7. Zherebtsov, I.P. "Elektromagnitnye kolebaniya i volny" v novom uchebnike fiziki "Electromagnetic Oscillations and Waves" in the New Physics Textbook). — In: Fizika v shkole, No. 1. 1955.

8. Ioffe, A.F. Fizika v srednei shkole (Secondary-School Physics). — In: Narodnoe obrazovanie. No. 3. 1958.

9. Kalashnikov, S.G. Elektrichestvo (Electricity). — GITTL. 1957.

10. Mackenzie, A.E. A Second Course of Electricity. — Cambridge. 1952.
11. Netushil, A. V. and K. M. Polivanov. Teoriya elektromagnitnogo polya (Theory of the Electromagnetic Field). 1956.
12. Presmann. A. S. Santimetrovye volny (Centimeter Microwaves). 1954.
13. Peryshkin. A. V. Kurs fiziki (A Physics Course). Part III. 1957.
14. Tamm, I. E. Osnovy teorii elektrichestva (Principles of the Theory of Electricity). 1957.
15. Telesnin, R. V. Elektrichestvo (Electricity). — GITTL. 1952.
16. Taucher. Physique. Paris. 1954.
17. Einstein, A. Zur Elektrodynamik bewegter Körper. — Annalen der Physik, No. 17. 1905.
18. Yus'kovich, V. F. and S. V. Roslavlev. Nekotorye itogi priemnykh ekzamenov po fizike (Some Results of the Entrance Examinations in Physics). — In: Fizika v shkole, No. 2. 1951.
19. Tsinger, A. V. Nachal'naya fizika (Elementary Physics). 1923.
20. Shakhmaev, N. M. Iz opyta oborudovaniya kabineta fiziki (Some Practical Findings in the Equipment of School Physics Facilities). — Uchpedgiz. 1957.

S. E. KAMENETSKII
Fellow of IMO APN*

THE USE OF ANALOGY IN THE SECONDARY-SCHOOL PHYSICS COURSE

INTRODUCTION

Every pedagogue knows from experience that an appropriate choice of analogies in the explanation of any given subject makes it much easier understood by the students.

Analogies have long been used, both in teachers' oral explanations and in physics texts. This methodological aid has been rather neglected lately, however. Thus, for instance, considerable use was made of analogies in the presentation of elementary electrical concepts in the standard textbook up to 1948, while from 1949 on they are missing altogether.**

On the other hand, in modern popular science publications, periodicals, and methodological literature abroad,† extensive use is made of analogies in the discussion of many difficult subjects.

Analogies seem to have fallen into disrepute to some extent as a natural reaction to the fact that a great many analogies used at the end of the 19th and the beginning of the 20th century were rather poorly chosen; this is partly due to the fact that analogies were more often than not aimed at a general simplification of the exposition rather than at making difficult problems easier to understand, and thus proved at times ineffective.

However, if the analogies are properly chosen, so that they uncover relationships between various phenomena and display the characteristic features of the phenomena, rather than just point to their superficial similarity, great possibilities are afforded to help the student correctly to grasp difficult aspects of the course. When analogies are contemplated in this light, the circumspection exercised in their use in the Soviet methodological literature, to the extent of their complete dismissal, cannot be justified by any means.

The effective and rational application of analogies is hampered by the lack of specialized publications dealing with the subject comprehensively enough. This served as a basis for conducting an investigation into the application of analogies in the secondary-school physics course, carried out by the author for four years (1954—1958), under the direction of his senior colleague, A. A. Pokrovskii (Institute of Teaching Methods). The

---

* [Institut metodov obucheniya Akademii pedagogicheskikh nauk (Institute of Teaching Methods of the Academy of Pedagogical Sciences.)]

** Peryshkin, A. V., G. I. Faleev, and V. V. Krauklis. Kurs fiziki (A Course in Physics), part II. 1949—1958.

† Cf. E. Dull. Modern Physics. —New York, 1951; Karl Geiger, Methodik der Lehre der Wechselstromtechnik. —Berlin, 1956; Fotyma, C. Z. and C. Z. Scislowski. Physik, Lehrbuch für die VII Klasse, Warszawa, 1953; etc. and others.

experimental teaching was conducted in two Moscow schools, Nos. 212 and 692. From time to time the physics teachers of schools Nos. 201, 215, 602, 689, and 717 (the Moscow Timiryazev district) were asked to participate in the work.

First, observations were performed on the teaching procedure so as to find out which topics were best suited to the use of analogies, the appropriate analogies were picked and worked out, and methods of their application were then tested. The students' progress was checked by means of written tests and oral discussions.

The question of models was also brought into consideration. As is known, two kinds of models are employed: physical models, in which the phenomena to be compared are of the same nature and differ only in scale (modeling in the restricted sense of the word), and mathematical models, where one phenomenon is represented by another, different in nature but described by the same equations as the original (modeling by similitude). The object of the investigation was to elucidate the pedagogical effectiveness of various analogies, mathematical models inclusive.

1. Analogy as a method of investigation and as an educational aid

Analogy serves different purposes in science and in education. Still, in order to place analogy as an educational aid, it is necessary to dwell, at least briefly, on its role in science.

The history of science testifies to the fact that analogies have materially contributed to the development of scientific thought. They have been effectively used by such great scientists as Maxwell, Thomson, Faraday, Hamilton, Hertz, Ohm, Lebedev, Stoletov, Umov, and others.

Maxwell enlarged upon Faraday's ideas and introduced into electromagnetic phenomena hydrodynamical images.

Analogies and models were used by Maxwell as a method of investigation and discovery; "by the use of this method Maxwell arrived at his famous equations."[*] Maxwell adopted the principle of the unity of nature, i.e., the similarity of the laws governing the behavior of diverse physical phenomena.

Another quite instructive and interesting instance of the application of analogy in science is the one that Ohm drew between an electric current and the flow of heat or water. Ohm assumed that when a voltage is applied to an electrical conductor, the flow of current through it is similar to that of a liquid under pressure through a pipe, or to the transfer of heat under the effect of a temperature gradient.

"Ohm was guided by such intuitive conceptions in establishing the famous law, which bears his name."[**] In this way he made it easier to vizualize such electrical concepts as current, voltage, resistance, and emf, which had remained unclear up to his time.

It is well known that the theory of electricity started evolving rapidly after Coulomb discovered the law of interaction of electrical charges. Coulomb "postulated that between any two electrical charges there acts a

---

[*] Kudryavtsev, P. S. Istoriya fiziki (History of Physics). Vol. II. p. 175.
[**] Ibid., Vol. I. p. 400.

force proportional to the respective charges, by analogy with Newton's law."* He also drew the analogy between electrical and magnetic phenomena and developed magnetostatics, by analogy with electrostatics.

The analogy between electrical and thermal phenomena (Thomson) is of particular historical interest. It sets up a correspondence between an electroconducting body and a thermal conductor, the electric potential at different points in a field and the temperature at different points in a body, etc.

Another fruitful analogy was the optical-mechanical, first employed by Hamilton in the consideration of the undular and corpuscular theory of light. Proceeding from another standpoint, Schrödinger made use of the optical-mechanical analogy to solve the problem of atomic energy radiation.

There is an interesting analogy between gases and solutes, which was first qualitatively formulated by Mendeleev and later developed by Van't Hoff.

A unified treatment of all cases of oscillations and waves, based on the extensive use of analogies both in theory and practice, was first propounded and then developed by L. I. Mandel'shtam, N. D. Papaleksi, and their pupils.

Analogies and models also find extensive use in modern physics. Recently, for instance, in his study of superconductivity, N. N. Bogolyubov established an analogy with superfluidity. Another very promising analogy is the one drawn between catalysts and semiconductors in the study of catalysis. A quite effective method of study is that of electrical analogues whereby "a mechanical or acoustical system is simulated by means of electrical circuits"**; the effectiveness of the method is due to the high degree of sophistication achieved by electrical engineering, especially in the analytical design of electrical circuitry.

From the realm of science, analogy gradually invaded the domain of education, as a useful teaching device. Scientists have intuitively resorted to it in expounding their ideas, as a means of explaining difficult problems. Finally, analogy made its way into the method of physics, and some physicist-methodologists advocated its use, as being not only of scientific but also of methodological value.

Out of the specialized investigations on the subject, two papers by V. L. Rozenberg are of especial interest: "Ob analogiyakh v nauke i prepodavanii"† (Analogy in Science and Teaching), and "Potentsiyal'naya teoriya elektricheskikh yavlenii na osnove analogii"†† (The Potential Theory of Electricity Based on Analogies). They present for the first time in a consistent manner the use of analogy in teaching, though most of the analogies proposed by Rozenberg have now become obsolete and are of little value.

The importance of analogy in teaching practice may be appreciated by examining the educational literature.

Analogies have been employed in the physics courses and textbooks of A. V. Tsinger, F. N. Indrikson, K. D. Kraevich, B. Yu. Kol'be, and A. I. Bachinskii, as well as in the methodology texts of E. Grimsel, N. B. Kashin, G. G. De—Metz, K. Hahn, F. N. Shvedov, and many others.

Part of the analogies used in the textbooks were borrowed from science. Such are, for instance, Thomson's analogy between electrical and thermal

---

* L a u e. M. History of Physics.

* O l s o n, G. Dynamic Similitude, p. 9. 1947.

† Journal "Russkaya shkola (The Russian School), Nos. III and IV. 1909.

†† Idem, Nos. 5, 6, 7, and 8. 1894.

phenomena (e.g., Rozenberg, Tsinger, Kraevich), Ohm's analogy between electric current and the flow of water (e.g., Indrikson, Bachinskii, Hahn, De-Metz).

Use was also made in the textbooks of analogies not found in the scientific literature, expressly worked out by physicist-methodologists for pedagogical purposes. In this respect particular imaginativeness was displayed by V.L. Rozenberg, J. Claude, A.I. Bachinskii, V.N. Bakushinskii, and some others.

A significant paper in this field is D.I. Sakharov's "An Aid for People Who Find Electricity 'Hard to Understand'", which largely draws on analogies for the explanation of the more difficult aspects. A broad range of apt applications of analogies may be found in J. Claude's book "Electricity and Its Applications Made Easy". The book went through quite a number of editions in France, and received a prize from the Paris Academy of Sciences; it was translated by W. Ostwald into German in 1908, and a Russian version of it appeared in 1910, revised and edited by the eminent proponent of analogies, A.A. Eichenval'd.

Mention should also be made of the application of analogies in university texts, although in this case there exist many alternative educational possibilities.

Of primary interest among the methodological texts are Kashin's "Metodika fiziki" (Methodology of Physics) and Hahn's "Methods of Teaching Secondary-School Physics", in which the authors advocate the use of analogies.

The application of analogies has been also given attention by contemporary methodologists. P.A. Znamenskii's "Metodika prepodavaniya fiziki v srednei shkole" (Methods of Teaching Physics in Secondary School) contains references to analogies, particularly in the study of electrical phenomena in the 7th grade. In his "Metodika prepodavaniya fiziki v srednei shkole" (Methods of Teaching Physics in Secondary School), I.I. Sokolov devotes a special section to analogies, in which he specifically advises against the definition of concepts through analogy. This opposition to the use of analogy as a means of introducing new concepts should not, however, be interpreted as a flat rejection of analogy in general. Thus the author says (p. 72, 1951 edition) that analogy "remains a good way of explaining and making familiar a new concept by comparing and relating it to other concepts."

The use of analogy is most explicitly and clearly discussed by E.N. Goryachkin, only as applied to electrical phenomena in the first stage of instruction, however. In his "Metodika prepodavaniya fiziki v semiletnei shkole" (Methods of Teaching Physics in the Seven-Year School), he prescribes methodological procedures which can be directly applied by teachers. Goryachkin, as well as D.D. Galanin,* derive in a straightforward manner a valid conclusion as to the effectiveness of using analogy in introducing new concepts in the elementary study of electrical phenomena.

Even a brief perusal of the methodological works and study books suffices to show the great importance of analogy in teaching. It is therefore hardly admissible that the role of analogies should be underrated as it is at present.

---

* Galanin, D.D. Ponyatie energii v kurse fiziki semiletnei shkoly (The Concept of Energy in the Physics Course of the Seven-Year School), pp. 48 and 55. 1947.

2. The pedagogical purposes of analogy

The necessity of making use of analogies in certain cases is conditioned by the complexity of the learning process. Some of the phenomena treated in the physics course are, in fact, difficult or altogether impossible to perceive by the senses (e.g., in electricity, part of optics, and many molecular processes). In such cases demonstrative experiments fail to display the intrinsic nature of the phenomena but only show their external manifestation.

It is, for instance, impossible to demonstrate the fact that the electric current in a conductor is a flow of electrons, as only the effects of the current are amenable to observation. Showing the operation of an electrical circuit and the readings of the instruments connected into it does not elicit the processes occurring in the circuit. A considerable amount of additional reasoning has to be done to associate the instrument readings with the actual processes taking place in the circuit.

Analogy proves an effective means of enhancing the intuitive appeal of teaching; it offers the students, whose powers of abstraction are limited, visual props which are quite valuable in helping them grasp the concepts they are being taught. For instance, in the formulation of the concept of electric current, the current is conveniently visualized by the use of images drawn from the flow of liquids.

The judicious choice of reference images facilitates the assimilation of subsequent concepts and helps quickly to secure positive results. These images are, of course, used only as a crutch and are discarded as soon as the concept has been properly formulated.

Analogies may in addition be employed to explain experiments which are of basic importance but which cannot be performed at school because of the experimental difficulties involved (e.g., Mandel'shtam and Papaleksi's experiment on the inertial motion of electrons, experiments designed to elicit the processes in direct-current and oscillatory circuits or tube oscillators).

It is also indubitable that the use of analogy in teaching, involving the analysis of the regularities prevailing in the phenomena under comparison, promotes the students' development. Setting up an analogy constitutes a logical operation, and it is just as desirable to be able to draw inferences by analogy as it is to apply induction and deduction.

The introduction of analogies is made possible by the fact that the students are already prepared for it. It has been conclusively shown[*] that the students have no difficulty in setting up a correlation between phenomena for the purpose of comparing them.

This capacity increases along with the students' development. But if the correlation and comparison of phenomena is to be made from the outset a means for understanding and memorizing concepts, it is necessary to teach the pupils to draw comparisons and to use analogies. The analogies have to be applied in a carefully thought out and systematized manner, not just at random. Their application is aimed at an easier and quicker assimilation of the taught material and at improving the students' knowledge. This can

---

[*] Smirnov, A.A. Psikhologiya zapominaniya (The Psychology of Memory), pp. 249, 264—280. 1948.

be achieved, because:

1) The intuitive appeal of the teaching is enhanced, as the students are enabled to visualize phenomena not amenable to sensory perception;

2) It becomes easier to elucidate the significance of physical processes and the laws governing them;

3) The reasoning power of the students is developed.

It ought to be pointed out that, as Sokolov correctly affirms,* analogy must not be used to define new concepts. However, when employed conjointly with other methodological aids, primarily experimentation, analogy proves quite a useful means of introducing some new concepts (e.g., voltage, potential).

3. The classification and limits of applicability of analogies

The teacher may be materially helped in the correct choice of analogies by a simple and convenient classification, which would take into account the specific requirements that the analogies have to meet in the physics course. The analogies are best subdivided according to three criteria, rather than one.

First of all, according to the way they are applied, analogies are divided into comparative analogies and analogies in the full sense of the word. The former are comparisons drawing upon similarity of behavior (e.g., the hydrodynamic analogy for the connection of conductors in series and in parallel) and are used in the first stage of instruction. The application of analogies in the full sense of the word involves considering a number of characteristics and drawing inferences (e.g., the analogy between thermionic emission and evaporation, between electric and gravitational fields). Such analogies are, of course, applicable only in the senior classes.

Secondly, according to the way in which inferences are drawn from them, analogies are divided into inductive and deductive. If one proceeds from the similarity of some characteristics in given objects in order to derive the similarity of some other characteristics, the analogy is termed inductive (e.g., the hydrodynamic analogies for electric current, the analogy between thermionic emission and evaporation); if, on the other hand, the conclusion is drawn, not from the similarity prevailing between some of the characteristics of two phenomena, but "on the basis of a common principle, a general assumption, from which the same consequences derive for both phenomena",** the analogy is then deductive.

For instance, in studying the phenomena of interference and diffraction in optics, it proves convenient to draw an analogy with the interference and diffraction of the mechanical waves on water. This analogy is based on the fact that light is an electromagnetic wave, and that any waves, irrespective of their nature, exhibit interference and diffraction. In this case it is assumed that one of the phenomena would manifest the same properties which are characteristic of the other, by virtue of the fact that both belong to the same class of phenomena. Analogies of this type are, for instance, the analogy applied to the study of resonance in electrical circuits, or the

---

* Op. Cit., §21.
** Strogovich, M.S. Logika (Logic), p.317. 1949.

analogy drawn between the processes occurring in a tube generator and in a self-oscillating mechanical system (e. g., a clock pendulum).

Thirdly, according to the place they occupy in the physics course, analogies may be labeled as follows:

1) hydrodynamic analogies employed in the study of electrical circuits;

2) the analogy between the electric and gravitational fields, applied to the study of the electric field;

3) analogies employed in the study of alternating current and electromagnetic oscillations;

4) analogies applied to the study of electromagnetic waves, optical phenomena inclusive;

5) analogies employed in the study of the structure of matter.

Any analogy proceeds, not from the mere identity of the phenomena involved, but rather from the similarity of some of their features, and it is therefore applicable only within certain limits. So long as these limits are not established, there is no point in trying to determine the scientifically or methodologically proper way of applying any analogy.

Now what defines the domain of application of an analogy, and how can its limits be recognized and complied with?

As a rule, an analogy cannot be carried further than the point at which the distinctive features of the studied phenomenon, which set it off from any other, come to the fore. Take, for instance, the hydrodynamic analogy employed in the study of electric current. This analogy helps to elucidate the concepts of current, voltage, resistance, and their mutual dependence. However, an electric current consists of a stream of charged particles, and it thus exhibits some effects (i.e., magnetic, chemical, and thermal) which cannot be interpreted in terms of any analogy, as these are properties inherent to electric current and not to any flow in general (e. g., of some given fluid).

In order to avoid inaccuracies in drawing up an analogy, whenever any two phenomena are compared it is necessary not only to dwell on the common points in their properties but also to bring out the intrinsic difference between them. It is best to point out these differences while working out the analogy, rather than all at once. The analogy is then developed up to the point when it proves possible to show up the difference between the phenomena. This is the way analogies have been applied in science, and that is also maintained in psychology, but the fact is often overlooked in teaching practice.

It is impossible to set limits of applicability which would hold good for every type of analogy. Each individual case requires specific analysis, which may be carried out only when the phenomena to be compared have been thoroughly investigated. This is readily achieved in the case of "methodological" analogies, as the phenomena brought under comparison have been already well investigated in the realm of science and their particular characteristics determined.

When analogy is used as a means of inquiry into some phenomenon still under scientific investigation, whose specific features are in the process of being elicited, its limits of applicability are hard to define and the analogy may turn out to be false or ineffective. Such analogies have happened in science. Newton, for instance, tried explaining the blue tinge of the sky by the presence of minute water droplets, drawing up an analogy with the colors produced when light passes through thin transparent plates.

Teaching, however, presents some special features. In the first place it makes possible, as already mentioned, to define the limits of application of any analogy, and furthermore it rules out borrowing false analogies from science. This is due to the fact that education lags to some extent behind research, so that methodology borrows only those ideas which have been tested in science. As a result teaching is more liable to be given to obsolete analogies rather than ineffective ones. Instances of such analogies are the analogy between electrical and magnetic phenomena (Coulomb), or the analogy between electrical and thermal phenomena (Thomson). These analogies were quite effective in their time, but as science evolved new features of the phenomena were brought to light which made the analogies outdated.

Thus, if the analogies are periodically revised so as to eliminate obsolete ones, and if they are applied in such a way as to demonstrate both the similarity and the difference in the phenomena under comparison, there are no grounds for rejecting their use. "There is nothing wrong with using analogies, only with their improper interpretation."*

4. The application of analogies in the elementary study of electrical phenomena in the 7th grade

Out of the well-known analogies used in the study of electrical phenomena, the one generally preferred** is the hydrodynamic analogy. This analogy is the most readily understood in the first phases of education. Moreover, the flow of water can be vividly demonstrated in the classroom, by means of fairly simple setups. The heat analogy has become obsolete, the gas analogy is more complicated and has less visual appeal, while the various mechanical analogies (e. g., the motion of small balls) oversimplify things and may create erroneous ideas.

Before the students proceed to the study of electric current, they are acquainted with the basic propositions of electrostatics. There, also, use could be made of analogies (A. V. Tsinger, E. N. Goryachkin, V. L. Rozenberg, etc.). It proves, however, of little value to employ analogies when introducing the basic ideas of charges, their interaction, etc., since the students readily grasp these notions through suitably chosen and performed experiments.†

When we come to the introduction of the concept of electric current, it is quite useful to make use of the anlagoy with the flow of water in pipes. This analogy becomes particularly descriptive if the concept of the electron is introduced and the electric current is represented as the orderly flow of electrons in the conductor.

The hydrodynamic analogy is quite effective in dealing with electric current sources. The point is that the term itself, "current source", gives the students a false idea. They conceive of the current source as

---

* Galanin, D. D. Op. cit., p. 55. 1947.

** Bachinskii, A. and S. Il'yashenko. Fizika (Physics), p. 328. 1952; Faucher, R. Physique, Classe de Premiere, p. 266. 1954, and some others.

† Cf. the experiments described in the book by Pokrovskii, A. A. et al. Demonstratsionnye opyty po fizike v VI—VII klassakh srednei shkoly (Demonstration Experiments in Physics for the 6th and 7th Grades of Secondary School), pp. 259—271. — Uchpedgiz. 1956.

some kind of device which "manufactures" electric charges that proceed to flow in the circuit. It is useful here to draw the analogy with a pump.* In a hydrodynamic system, the pump does not actually produce water but only causes it to move along. The current source acts similarly in an electric circuit; it does not produce charges but only sets into motion the charges present in the conductors.

After the students have been acquainted with the different components of an electric circuit, a simple electrical circuit is connected up, consisting of a current source, a consumer, connecting wires and a switch. A hydrodynamic analogue of the circuit is also assembled.

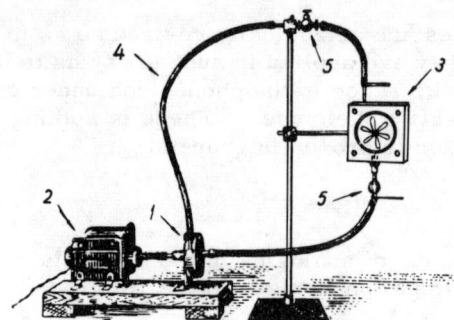

FIGURE 1

The hydrodynamic analogy is demonstrated on a setup (Figure 1) made up of a centrifugal pump — 1, an electric motor — 2, a water turbine — 3, connecting rubber tubes — 4, and stopcocks — 5.

The electric current source is likened to the pump, the electric energy consumer to the water turbine, the connecting wires to the water-filled tubes, and the switch to a stopcock.

The centrifugal pump used in the setup has a hermetically sealed body with two connecting sleeves (intake and supply pipes), onto which are slipped the rubber tubes. The shaft of the vane wheel of the pump is lodged in a gasket and is connected to the shaft of the motor.**

The speed of rotation of the motor can be controlled by means of a rheostat. The water flows into the turbine through a nozzle at the top, sets into motion the rotor, and flows out through an outlet at the bottom.

The centrifugal pump and the motor are fastened to a common support and the water turbine is fixed onto a laboratory stand. The system is filled up with water by means of a siphon; in order to let the air out during the filling, the tubes at the top of the setup are pulled apart.

The components of this simple setup work on a principle with which the students are already familiar; besides, their mode of action is easily demonstrated. Thus, the operation of the centrifugal pump is simply demonstrated by showing the jet shooting out of the supply pipe. The way in which the turbine works is readily understood, as it is made out of plexiglas and its parts are exposed to view.

* Cf. Bachinskii, Op. cit., p.331. 1952.
** In the setup described use was made of a 75-watt electric motor from a sewing machine.

After demonstrating side by side the operation of the electrical circuit and of its hydrodynamic analogue, the two setups are schematically drawn on the blackboard (Figure 2). It is then explained to the students that electric charges circulate in the functioning electrical circuit, just as the water does in the hydrodynamic system. If the circuit is broken at any point, the circulation is disrupted. This explanation makes it possible to eliminate a mistake the students commonly make, of claiming that the switch has to be necessarily placed between the positive pole of the current source and the consumer, i.e., before the consumer, following the direction of the current in the circuit.

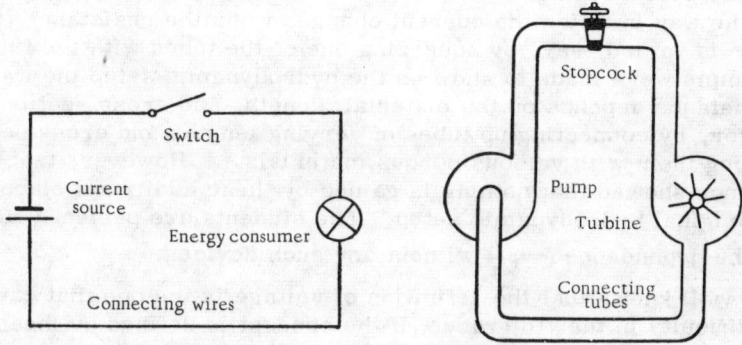

FIGURE 2

The hydrodynamic analogy also provides the means of explaining what the current intensity means, and the fact that the current is constant everywhere in the circuit so long as it does not branch out at any point. After the construction of the ammeter has been explained and its operation demonstrated, the students have no trouble in understanding that the ammeter has to be connected into the circuit in series. The hydrodynamic setup is analogously supplied with a flow meter, whose readings simulate those of the ammeter (Figure 3).

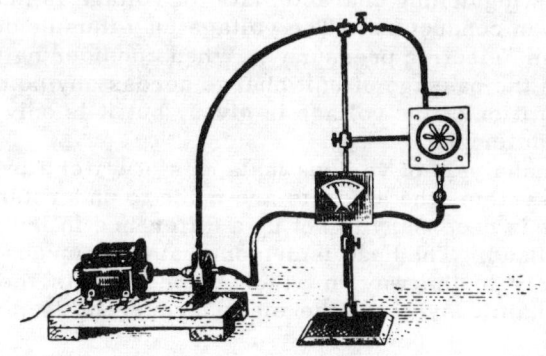

FIGURE 3

The flow meter has a channel running through it, with a plate set perpendicular to the water flow and fixed to the pointer. The pressure of the moving water on the plate causes the pointer to be deflected. In order that the readings should depend on the flow velocity, a coil spring producing an opposing torque is attached to the pointer. The flow meter is made out of plexiglas and its working is exposed to view.

The connection of conductors in series and in parallel is explained by means of examples with which the students are familiar from everyday life (e.g., streams, channels). The concept of resistance is first presented with the aid of pictorial similes (e.g., the movement of people over a bridge, a path, or over rough ground).* Then use is made of the setup to demonstrate the way in which the current changes when the resistance to the flow of water is varied, say, by squeezing one of the tubes with the fingers.

Attempts were made to show on the hydrodynamic setup the way in which the resistance depends on the material, length, and cross section of the conductor, by connecting up tubes of varying lengths and cross sections and filling them with various porous materials.** However, teaching experience showed that nothing is gained by these additional elaborations of the original hydrodynamic setup. The students are perfectly able to grasp the dependence $R = \rho \frac{l}{s}$ without any such devices.

It is well known that the definition of voltage is the one that causes the most difficulty in the 7th grade. If the concept is defined by means of analogy there is some loss in rigor, while if the definition proceeds from considerations of energy considerable difficulties arise as 7th graders find the concept of energy quite hard to understand. Repeated attempts were made to combine these two approaches.† Our method of defining the voltage is, in fact, based on a synthesis of analogy, experiment, and energy considerations.

It should be borne in mind that at the beginning reference is only made to the voltage of the current source. Indeed, the students are first acquainted with voltage in the consideration of the various kinds of current sources, when they are told that electrification occurs in an electric cell and that one of the terminals is taken as positively charged, and the other one as negatively charged. The voltage across a section of the circuit is defined much later.

The chief distinguishing characteristic of voltage is that it produces a flow of current in conductors. The voltage may thus be defined in the 7th grade only as an "electric pressure". When considering the energy associated with the passage of unit charge across any point in the circuit, though, no definition of the voltage is given, but it is only shown how the voltage is characterized.

Through the analysis of various instances of water flow and of a closed hydrodynamic system, the students are made to understand that for the water to flow it is necessary to set up a difference in levels, a pressure differential or head. The head is demonstrated by means of the centrifugal pump. Similarly to the way in which the head sets the water flowing in the hydrodynamic system, the electric current is produced in the

---

* Op. cit., pp. 344-336.

** Sakharov, D.I. V pomoshch tem, kto "plokho ponimaet" elektrichestvo (An Aid for Those Who Find Electricity "Hard to Understand"), p. 24. 1931.

† Goryachkin, E.N. Metodika prepodavaniya fiziki v semiletnei shkole (Methods of Teaching Physics in the Seven-Year School), Vol. 1, p. 382. 1948.

electrical circuit under the effect of the "electric pressure"* which we call voltage.

When explaining the role of the voltage in an electrical circuit, the teacher shows that a light bulb connected to a dead storage battery does not burn, while if connected to a charged battery, across whose terminals there is a voltage potential, the bulb lights up, showing that current is flowing through it.

It is next necessary to study the concept at further length, and clarify how the voltage is characterized.

It is a known fact that flowing water performs work. Electric current performs work in a similar way. The voltage may be assessed from the amount of work performed. In this case, however, the dependence is more involved, as the work has to be expressed not in terms of the whole charge flowing through but in terms of the passage of one coulomb. The voltage is characterized by the work performed by unit quantity of electricity flowing across any given section of the circuit.

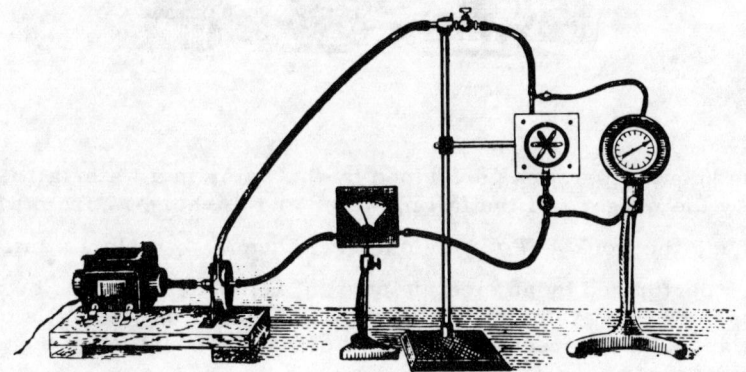

FIGURE 4

It is quite useful to apply the analogy of a voltmeter with a manometer. To this end a demonstration is made with the hydrodynamic setup fitted with a manometer whose readings simulate the readings of the voltmeter (Figure 4). The manometer has a high-response aneroid box, capable of reacting to fairly low pressures (a few millimeters of mercury). Its operation may be first demonstrated by connecting it to a vessel filled with water, and showing how the manometer readings change with any small change in the height of the vessel. The manometer is then connected to the turbine and shows the pressure differential across it.

An electrical circuit is hooked up in a similar way (Figure 5). These two analogs are compared to show the behavior of the voltage in circuit. It is also shown that like the manometer, the voltmeter has to be connected in parallel to the section over which the voltage is to be measured, and that it must have a high resistance so that little current should be diverted into it.

* Bachinskii, A. and S. Il'yashenko. Op. cit., p.347.

Having reached the point where all the basic quantities (I, U and R) have been defined, Ohm's law for any section of the circuit can be easily derived by experiment,* without the use of analogy. It is also a simple matter to introduce the formulas expressing the work and the output, though for the purpose of illustration it may be a good idea to draw the analogy with a waterfall.**

FIGURE 5

As is known, the work performed by the water in a waterfall is determined by the weight P of the falling water and the height h, from which it falls, i.e., the work $A = P \cdot h$; the output is then $N = \frac{Ph}{t}$. In a similar manner, the work performed by an electric current is $A = qU$, while the output $N = I \cdot U$. Working out some examples of the work and output in waterfalls of different heights and with different rates of flow of water should make these relationships quite clear.

Thus, the hydrodynamic analogy may be employed to study electrical circuits and to define the basic concepts of current, voltage, resistance, work and output. When proceeding to the study of the magnetic properties of electric current, the question as to a hydrodynamic analogy may occur to the students. They should be given to understand, however, that the analogy is no longer applicable in this case.

5. The application of analogies in the study of electrostatics in the 10th grade

It is well known that the greatest difficulties are encountered in electrostatics in defining the electric field, potential, and electric capacitance.

Two analogies have been commonly adopted in the study of the electric field, viz., the analgoy between electrical and thermal phenomena, and the

---

* Cf. the experiment described in the book by Pokrovskii, A.A., et al., op. cit., pp. 227—228.
** Putilov, K.A., in conjunction with V.A. Fabricant. Kurs fiziki (A Course in Physics), Part II. p. 68. 1945.

analogy between the electric fields and the gravitational field. The first one is not very helpful in understanding the idea of the electric field and of its characteristics, because no parallel is drawn between the different fields. In the study of the electric field in the 10th grade it is more practical to make use of the analogy between the electric field and the gravitational field. This analogy was first applied in a consistent manner by A. I. Bachinskii.*

The analogy is based on the fact that the two fields are derived from a potential and that it is possible to calculate the work done on moving a charge q and a body of mass m in either field by similar formulas, i.e.,

$$A = m(gh_2 - gh_1) \text{ and } A = q(\varphi_2 - \varphi_1).$$

By the use of this analogy the study of new material is organically linked with reviewing of the old material.** Thus, in the study of electrostatics such questions are reviewed as the law of universal gravitation; the independence of the work of the form of the path in a gravitational field; the potential energy; the units of force and work, etc.

The whole theory of electrostatics could be constructed by analogy with the theory of gravitation, but such a presentation is impractical in secondary school. Analogies must be used only where their application is dictated by methodological considerations.

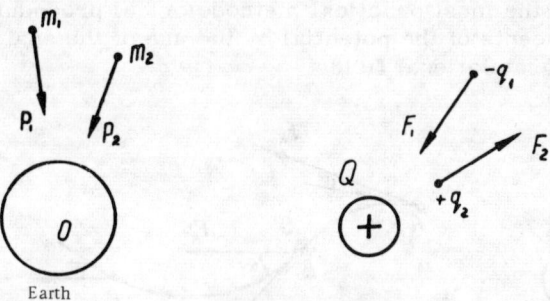

FIGURE 6

The analogy between the gravitational and the electrical fields is employed primarily in the definition of the electrical field, this concept being difficult to grasp as it cannot be directly perceived by the senses. In this case experiments provide the means only of demonstrating the interaction between electrical charges. The analogy with the gravitational field saves the discussion from empty formalism.

Drawings are made on the blackboard (Figure 6).

The explanation provided in this case runs basically as follows:

Throughout the space surrounding the earth forces act on any body. The action of these forces is due to the presence of a particular kind of field, i.e., the gravitational field, which is the cause for the weight of bodies. The gravitational forces are always forces of attraction.

Throughout the space surrounding an electrical charge forces act. The action of these forces is due to a particular kind of field, i.e., the electrical field. The electrical forces may be either forces of attraction or of repulsion.

* Bachinskii, A.I. Fizika v trekh knigakh (Three Books in Physics), Book III, p.51. 1928.
** Rymkevich, P.A. Povtorenie kursa fiziki v X klasse (Reviewing the Physics Course in the 10th Grade), p.85. 1957.

This analogy unfortunately provides only a geometric image of the field. But even a geometrical representation may at first be useful, while subsequently the ideas are elaborated, it is proved that the electrical field is a material entity, and the content of the concept is expanded.

It is also a matter of course to make use of the analogy between Coulomb's law $F=\frac{q_1 \cdot q_2}{R^2}$ and Newton's law $F=\gamma\frac{m_1 \cdot m_2}{R^2}$. This analogy is just formally applied and is valuable only for memorizing Coulomb's law.

It may appear quite useful, at first sight, to appeal to analogy in studying the electric field intensities.*

The analogy in question draws a comparison between E and g, but the students have quite different conceptions of these two quantities. In fact, $g=\frac{P}{m}$ is the acceleration of the force of gravity and is studied in kinematics. The force acting on unit mass in a gravitational field is something new for the student. On the other hand, E denotes the force acting on unit charge at a given point in an electrical field. Thus, the analogy between E and g is only liable to confuse the students and is of doubtful value.

As is known, in addition to its intensity, the field is also defined by means of the potential. It is very important that the students should be made to understand the reason for which this concept is introduced, and in what way it applies to the electrical field. Often these questions are not brought up and the potential is introduced in a formal manner, and consequently it is poorly understood.

Let us consider the most practical methodological procedure for introducing the concepts of the potential by the use of the analogy between the electrical and gravitational fields.

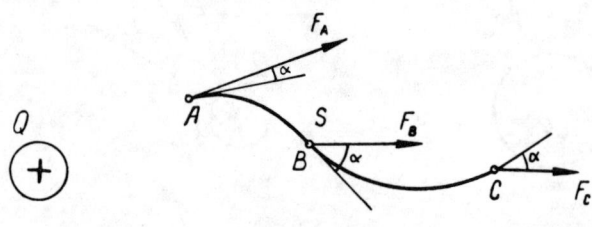

FIGURE 7

1. It is necessary to show that some difficulties are involved in calculating the work done on moving an electrical charge in an electrical field; the potential is introduced with the intent of overcoming these difficulties. Suppose it is necessary to calculate the work done in moving an electrical charge q in the electrical field produced by a point charge Q, between the points A and B over the path s (Figure 7). The usual way of calculating the work by the formula $A=F \cdot s \cdot \cos\alpha$ is in this case inapplicable, because F and α are not constant. It is, however, possible to make use of a special property of the electrical field, which enables the concept of the potential to be introduced, considerably simplifying the calculation of the work.

---

* R y m k e v i c h, P. A. Povtorenie kursa fiziki v X klasse (Reviewing the Physics Course in the 10th Grade), p. 85, 1957.

2. To this end, let us first consider how we calculate the work done on moving a body in a gravitational field. If a body of weight P is moved between the points A and B over the path s (Figure 8), then $A = F \cdot s \cdot \cos \alpha$. However, due to the fact that the work in a gravitational field does not depend on the form of the path over which the body is moved, and depends only on the points between which the motion is performed, it is possible to define the work as the change in potential energy of the body (not taking into account any resistance to that motion), i.e.,

$$A = W_{\text{pot}_B} - W_{\text{pot}_A} = Ph_2 - Ph_1 = m(gh_2 - gh_1).$$

It ought to be enough to remind the students of this property of the gravitational field, as they should be quite familiar with it.*

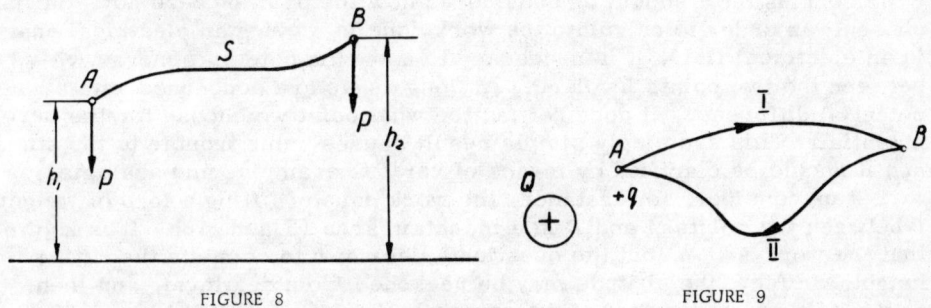

FIGURE 8   FIGURE 9

3. We shall prove that the electrical field, like the gravitational field, is a potential field, i.e., the work done in it does not depend on the form of the path of displacement. The fact that the electrical field is derived from a potential may be first fairly easily proved for the case of a homogeneous electrical field, and then generalized to the case of a nonhomogeneous field. The proof may also be carried out for the general case. Suppose that when a charge q is moved between the points A and B (Figure 9) in the field of a point charge Q, the work A_1 is performed over the path I and the work A_2 is performed over the path II. Now, let the charge move under the action of the electric field from the point A to the point along the path I, and let us then move it back to the point A against the field force, but performing the work A_2. If we now assume that $A_1 > A_2$, we should get as a result perpetual motion. The same situation obtains if we assume that $A_2 > A_1$, only reversing the direction of the circuit. The only possible assumption is that $A_1 = A_2$, which proves the potential nature of the electrical field.**

---

* This is considered in the case of an inclined plane in the: "Kurs fiziki" (A Physics Course) of A.A. Beryukin and V.V. Krauklis, part I, p. 134, as well as in other problems.
** Ioffe, A.F. Kurs fiziki (A Physics Course), Vol.I., p. 127. 1927; Rymkevich, P.A. Povtorenie kursa fiziki v X klasse (Reviewing the Physics Course in the 10th Grade), p. 84, 1957.

4. Let us now introduce the concept of the potential of the electrical field. As in the case of the gravitational field, which may be characterized by the magnitude of the potential energy of unit mass situated at any given point in the field, it is possible to introduce a special characteristic in the electrical field, viz., the potential, equal to the magnitude of the potential energy of unit positive charge situated at any given point in the field ($\varphi = \frac{W_{pot}}{q}$).* The work done on moving unit charge between two points in the electrical field is then given by means of $A = \varphi_2 - \varphi_1$ and in the case of an arbitrary charge q, by means of $A = q(\varphi_2 - \varphi_1)$. It is desirable to define the potential as $\varphi = \frac{W_{pot}}{q}$ and not as $\varphi = \frac{A}{q}$ even though $W_{pot} = A$, since the potential energy is equal to the work done on moving the charge to a given point in the field from the point where the potential energy is taken as zero.

In conclusion it should be considered how the point of zero potential is chosen. In order to calculate the work done in moving an electrical charge in an electrical field, it is necessary to know the potential energy ($\varphi_2 - \varphi_1$) between the two points involved. As long as we are concerned with a potential difference, it does not matter what point we choose for the zero potential. This seemingly simple result causes some trouble to the students, and it should be clarified by means of various examples and analogies.

Let us consider, for instance, the work done of lifting a load of weight P between the points A and B in a mountain area (Figure 10). It is known that the work $A = P \cdot h$, but the question arises how to compute the lifting height. In fact, the altitude may be reckoned from sea level, and then $h = h_2 - h_1$ or from the foot of the mountain, and then $h = h_2^1 - h_1^1$, or else from the level of the point A, in which case, the height is equal to h. However, as is known, $A = P(h_2 - h_1) = P(h_2^1 - h_1^1) = P \cdot h$, i.e., only the difference in altitudes has to be given for calculating the work, while the altitude itself may be reckoned from any given level.

The same holds for the work in an electrical field. The zero potential can be taken as the potential of an arbitrary body at an arbitrary point in the field.

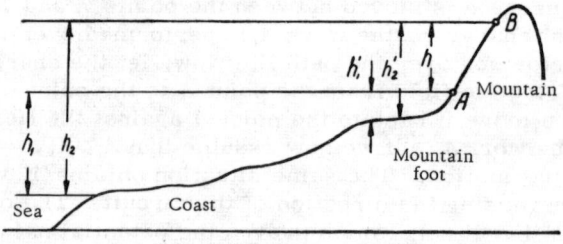

FIGURE 10

As may be seen, the potential is not defined in this case by analogy with the potential of the gravitational field; it is expedient however, to define this concept through the use of the common laws governing the two kinds of phenomena. It is first shown by experiment how electrical charges interact

---

* [The symbol used in the Russian text has been retained here even though the letter V is the symbol more commonly used for potential.]

with each other, and the fact that electrical field possesses energy, and throughout the subsequent discussion use is made of the concept of energy. Proceeding from the common properties of the electrical and gravitational fields, it is shown how the electrical field can be described in terms of the energy by means of the potential.

Analogy can be also useful in the study of electrical capacitance. This concept ($C=\frac{q}{\varphi}$) is introduced on the basis of a demonstration experiment and seems to be quite clear to the students. There are, however, some aspects which require further explanation. First of all, the student tends to associate the term "capacitance" with the idea of "capacity," or available space. It is necessary to show the difference between the concepts. The capacity of a vessel denotes the amount (the mass or volume) of matter which can be contained in the vessel. When reference is made to the electrical capacitance, the same idea occurs at first. For instance, the electrical capacitance of a capacitor is considered like this: "a capacitor contains a certain electrical charge in the same way as a vessel contains a certain amount of liquid."* In order to dispel any incorrect notions, use has to be made of an analogy.

Consider a pail. The capacity of the pail is finite, i.e., it can contain a definite amount of water. On the other hand, the question may arise as to the charge which may be contained in a body. This question may appear strange to anybody who has correctly understood the concept of capacitance, but it seems to cause trouble to the students, as they quite often give wrong answers to this question. The problem may be subsumed as follows (Figure 11):

A pail can contain only a definite, limited amount of matter.
A body can contain any given charge, only in this case the potential of the body varies.

In addition, it is not always clear to the student why the capacitance depends on the charge and the potential when it is not a property of either of them but a property of the body, i.e., they fail to understand in what indirect way q and φ determine C.

Drawing an analogy between the electrical capacitance and the capacity of a gas bottle is helpful in this case. It is known that the quantity of gas contained in the bottle is not enough by itself to determine the capacity since the gas may be under different pressures (Figure 12).

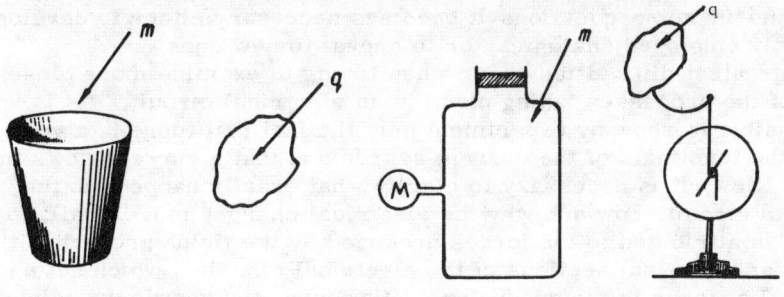

FIGURE 11   FIGURE 12

* Blonskii, A.F. Radio (Radio), p.18. 1957.

1. If a certain amount of gas m is contained in the bottle, the pressure will be p (with $T=$const).	If a charge q is imparted to the body, the potential of the body will be φ.
2. If the bottle should contain an amount $2m$ of gas, the pressure will be $2p$, for 3 it will be $3p$, etc.	2. If a charge $2q$ is imparted to the body, the potential increases to 2φ, for $3q$ it grows to 3φ etc.
3. The capacity of the bottle is defined, not by m, but by the ratio $$\frac{m}{p} = \text{const.}$$	3. The capacitance of the body is defined, not by q, but by the ratio $$\frac{q}{\varphi} = \text{const.}$$

It may also be made clear why in order to determine the capacitance one does not take simply the dimension of the body. Thus, for instance, if the bottle itself is not visible, its volume cannot be determined by direct measurement, but it may be determined indirectly through the ratio between the mass of the gas contained in it and its density. The same holds for the electrical capacitance. We have no way of measuring capacitance directly. It has to be measured indirectly, by means of the ratio $\frac{q}{\varphi}$.

Analogies are convenient to use only when considering the above problems. In all other problems (e.g., the dependence of the capacitance on the surface of the body, on the presence of neighboring bodies, on the medium) analogies are unnecessary, also considering the fact that in this case the analogies become involved and are of little help. It is possible to use analogies when discussing capacitors, but they are more conveniently employed in the study of alternating-current circuits into which a capacitance is connected.

6. The application of analogies in the study of constant current in the 10th grade

In the present case, analogies are employed mainly in the study of current in metals. When current in liquids and gases is considered, analogies prove practically unnecessary, since it is possible to make use of some previously learned concepts.

Analogies may be primarily employed to remind the student of the basic concepts studied in the 7th grade. This kind of analogy is already known to us.*

In the 10th grade, however, electrical phenomena are studied in more detail, and in some questions it becomes necessary either to develop previously employed analogies, or to appeal to new ones.

The greatest difficulties occur when trying to examine more closely the nature of the processes taking place in an electrical circuit. As is known, it is possible to show by experiment only the fact that there is a voltage across the terminals of the current source and that a current flows in the circuit. Now, it is necessary to clarify what exactly happens in the electrical circuit, how and why the electrical charges move at all. This process is attributed to the forces produced by the fields present in the internal and external sections of the electrical circuits, which act on the electrical charges in the conductors. However, the knowledge gained from any explanation is materially improved if the explanation is visually interpreted.

In order to explain the forces acting in an electrical circuit and to illustrate the potential jump, Professor Michelson employed in his time a

* See section 4.

visual aid in the form of a spiral inclined plane, pointing out that "at the place where an electromotive force acts in the circuit, the electrons are 'lifted' from a lower to a higher potential."*

In this instance, the point consists essentially in simulating the motion of charges in an electrical circuit by the motion of a body in a gravitational field. Most currently employed physics textbooks also appeal to such visual devices to explain the action of external forces and to exemplify potential jumps.

Experience has shown that it is quite helpful to make use of a demonstration model in the form of a spiral inclined plane down which a small ball is made to roll. The student should be able to build such a model himself.

It is best to take for the base of the model a plywood disc about 25 cm in diameter. The track for the ball is made out of two parallel lengths of curved wire; the wires are joined by concave supports, which are fixed to four wooden props set on the edge of the disc (Figure 13). The height of the props is adjusted so that the ball should drop about 10 cm.

The same principle may be more conveniently used to make a model out of thin sheet metal. A sheet is cut out in the shape of a trapeze, nailed to the rim of the disc and soldered where its two edges meet. Antoher sheet is similarly cut, shaped and placed inside the first one so that the two cut edges form a track for the ball to roll over. Between the metal sheets wooden props are fixed (Figure 14).

This model may be fitted with a lifting mechanism, which is separately assembled (Figure 15) and fixed inside the model with a bolt (Figure 16). It consists basically of a clockwork mechanism (without the escapement), 1, moving at a uniform rate. The spindle of the mechanism carries a revolving drum, 2. Another drum is fixed to a support attached to the clockwork mechanism. A rubber band, 3, having on it two ribs, 4, is wound over the drum. The clockwork mechanism is driven by a weight, 5. A small platform, 6, serving to lift the ball, slides freely over two guide wires, 7, and is moved up by each rib on the rubber band. The platform is cut out of sheet metal and has grooves to keep the ball in place. When the platform rises to the top the ball rolls off it, and then the platform is released from the rib on the belt and drops down. The ball rolls onto the platform at the bottom after having rolled down the incline. The next rib then lifts up the platform with the ball, and so on. The model is set on a ring fixed to a universal stand.

The demonstration is preceded by a discussion of the conditions necessary for the existence of a current in the electrical circuit (the presence of free electrical charges and of an electric field) and it is proved that the circuit must include a current source across whose terminals a constant potential difference is maintained.

The students are shown on the mechanical model how the ball rolls down under the effect of gravity. This motion is similar to the displacement of the electrical charges over the external electrical circuit under the effect of the electrical field forces. The ball is then lifted by hand, and work is thus performed against the force of gravity, similar to the work of the non-electrical forces in the electric current source. If use is made of the model with the clockwork mechanism, the ball will keep being lifted and rolling off until all the available energy driving the lifting mechanism is

---

* Michelson, V.A. Physics, Vol.2, p.73. 1918; pp.106 and 120. 1940.

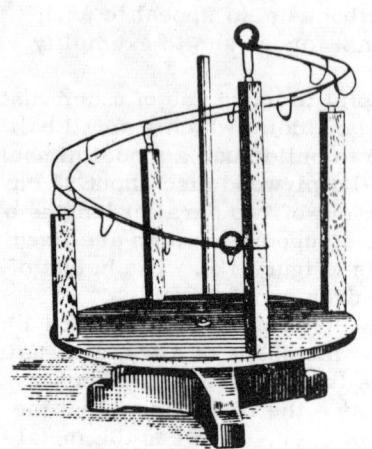

FIGURE 13

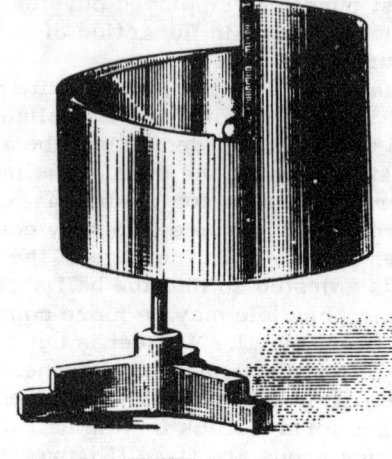

FIGURE 14

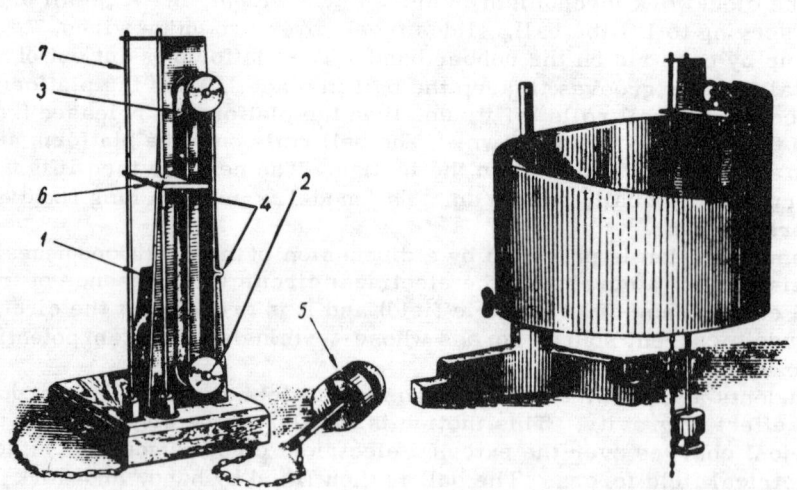

FIGURE 15

FIGURE 16

expended. Thus, it illustrates the necessity of continuously acting non-electrical forces, and it stresses, in addition, the fact that the current source will function only until its energy reserve is spent.

The nature of electric current in metals is usually explained on the basis of L.I. Mandel'shtam and N.D. Papaleksi's experiments (1912) on the inertial motion of electrons. The basic idea of this experiment may be vividly demonstrated on a mechanical model simulating the inertial motion of electrons.*

FIGURE 17

The model consists of a ring-shaped glass tube filled with water and some kind of solid particles (e.g., beads, grain, tea leaves) (Figure 17). If the tube is now made to rotate rapidly by means of a centrifuge and then stopped, the liquid in it will keep moving by inertia, and this will be made apparent by the solid particles. The electrons move in a similar way, when the coil is suddenly stopped, in the experiment mentioned above.

It is not at all clear to students, and, indeed, they have trouble coping with the fact, that, given the small velocity of motion of the electrons, there should be such a huge "current velocity" (about 300,000 km/sec) in conductors. There is no way of performing any demonstration to show this. The most appropriate way of explaining the fact is to draw an analogy with an oil or gas pipe,** thus comparing the "current velocity" with the velocity of propagation of pressure in the oil pipe, and the electron velocity with the velocity of motion of the oil particles themselves.

It is, however, much better to make use in this case of the simple hydrodynamic model, which may be easily put together in school. The model is made up of a vessel containing dyed water, with a long, horizontally

---

* Zenchuk, P.S. Elementy elektronnoi teorii v kurse elektrichestva v X klasse srednei shkoly (Elements of Electron Theory for the Electricity Course in the 10th Grade of Secondary School), M.Sc. thesis, p. 230. 1951.

placed transparent tube connected to its lower part; the tube is filled with clear water and fitted with a stopcock (Figure 18). If the cock is opened, the water immediately starts pouring out of the tube, and the boundary of dyed water slowly moves down the tube and reaches its end after a certain length of time. The motion of dyed water is thus made to simulate the motion of electrons in the electrical circuit.

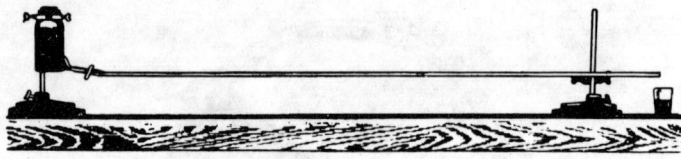

FIGURE 18

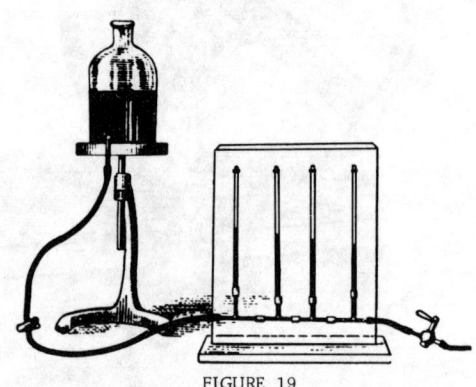

FIGURE 19

The most difficult concept in this particular case proves to be the emf. This concept may be made clear by appealing again to the mechanical model previously described. In the present case, the emf is analogous to the work done on lifting the ball against the force of gravity. Experience shows that the students find it hard to understand why the definition of the emf refers to the work of unit charge ($E = \frac{A}{q}$) over the whole closed circuit, while the work involved is performed by the non-electrical forces only over the internal section of the circuit. In the secondary school it is enough to give the students a purely qualitative interpretation, which consists in explaining that the work of the non-electrical forces contributes energy to the circuit, which is then expended over the whole circuit.

It is often recommended to illustrate the concept of the emf by means of various analogies (e.g., sliding down a hill on a sleigh, lifting a load up a hill); at the outset, however, it is more desirable to demonstrate a model in class.

* Peryshkin, A.V. Kurs fiziki (A Physics Course), Vol. III, p. 47. 1957.

It is very important to draw the distinction between the emf and the voltage across the terminals of the current source. It is quite useful to do this by means of a hydrodynamical analogy, where the emf is simulated by the pressure difference produced by a pump when water is not flowing in the system, and the voltage is simulated by the pressure difference when there is flow of water. All this can be demonstrated on a working setup. A manometer is used to show the pressure when the stopcock is opened and the turbine is rotating. The manometer readings represent in this case the voltage. The stopcock next to the turbine is then closed, the turbine stops, and the manometer readings increase. In that case, the manometer readings represent the emf.

It should be noted that the demonstration model consisting of communicating vessels (Figure 19), which is often employed to illustrate the voltage drop, is not very practical, because the students do not associate it in their minds with a closed electrical circuit.

When studying thermionic emission it is a good idea to draw the analogy with evaporation, which helps understanding the concepts of the work function and space charge.

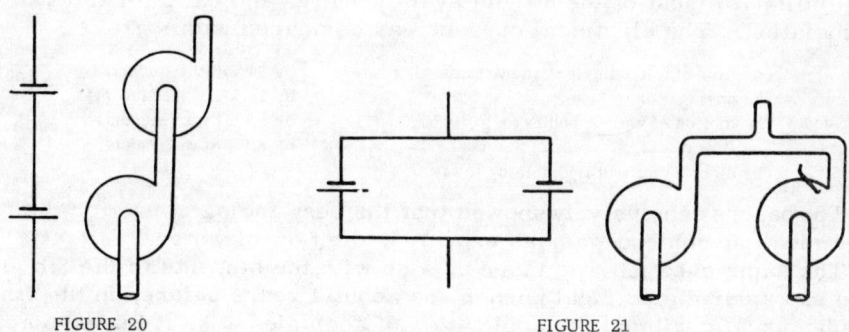

FIGURE 20   FIGURE 21

When considering the ways in which cells are connected in batteries there is a point in resuming the analogy with pumps, first employed back in the 7th grade. This analogy (Figure 20) has a visual appeal and illustrates well the fact that wehn the cells are connected in series the emf increases, in the same way as the pressure increases when the pumps are connected in series. When the cells are connected in parallel, the emf of the battery remains the same as the emf of a single cell, just like the pressure does not increase when the pumps are connected in parallel (Figure 21).

7. Experimental methods and some of their results

The basic propositions, conclusions, and methodological procedures presented above were for the most part tested by experimental work. Depending on the given conditions, various methods were employed: test

---

* Peryshkin, A. V. Kurs fiziki (A Physics Course), Vol. III, p. 47. 1957.

papers, individual discussions with students, observations in the classroom, interrogation of the students in class, etc. Let us adduce some concrete examples.

One of the primary questions arising in the investigation was the choice of the most practical analogs to be used in some specific topics of the physics course. In order to find an answer, papers were given to be written in the form of questionnaires.

Thus, in order to select an analog for electrical current, 7th grade students were given a questionnaire to fill out, before having studied electrical phenomena. The students were informed that the papers would not be graded, and that they did not have to be signed. In order to rouse the students' interest, they were told the purpose of the paper, viz., to establish their knowledge and ideas about electrical phenomena, so as to find the best way of presenting the relevant material in their subsequent studies.

The students were posed the question as to what they thought happened in the wires when the light is switched on when coming into a room, and whether they knew of any similar phenomena.

It was found from the answers of 250 students that the term "current" is familiar to most of them from everyday life, and calls forth definite association. The electrical current was compared with:

The flow of water in pipes and streams	by 44% of the students
The propagation of heat	by 15.5% of the students
The motion of various bodies (e.g., balls, sand)	by 9.5% of the students
The flow of gas	by 8.0% of the students

The other 23% could find nothing to say.

The papers conclusively showed that the best analog to use for the electrical phenomena was, as expected, the flow of water.

The same question was taken up also with the students of the 8th grade, who had studied electrical phenomena about 2 years before, in the 7th grade, usually without the application of analogies. As it turned out, most of these students also compared electrical current with the flow of water.

A similar procedure was adopted to find the best analog to use in all other cases as well. Thus, for instance, for the study of oscillation the choice fell on a spring pendulum, rather than on a mathematical pendulum; for the study of wave motion, surface waves in water were chosen, rather than sound waves.

When applying specific analogies in teaching practice, it is necessary to check whether this helps learning the new material, and whether the knowledge gained is in any way improved in extent and content. To this end, the same material was explained to different student groups, in an identical way as far as the scope, order of presentation, and details were concerned, but in one of the groups the teaching was supplemented by analogies. The results were checked by means of test papers. The following were thus checked: the simulation of a current source by a pump, the hydrodynamic analog of an electrical circuit, the analogy between the electrical field and the gravitational field, etc.

However, in order to find an optimal solution, it was not enough simply to supply analogies, but also to introduce variations in the presentation of the material and to use various methods for checking the results. Let us consider as an example the location of the switch in an electrical circuit.

It is a fact that students in the 7th grade often mistakenly think that the switch in an electrical circuit must be placed between the positive terminal of the current source and the consumer and not just anywhere in the circuit. This mistake is further complicated by the fact that electrical circuits are studied immediately following the discussion of the direction of the current. By changing a little the order of exposition, introducing the direction of the electrical current after having discussed electrical circuits, and incorporating in the laboratory work the hydrodynamic analog which illustrates the circulation of the charges in the electrical circuit, a better result was secured, which was shown by a new test paper.

It was further necessary to show that the problem cannot be satisfactorily solved by demonstrations and laboratory work without the use of analogs.

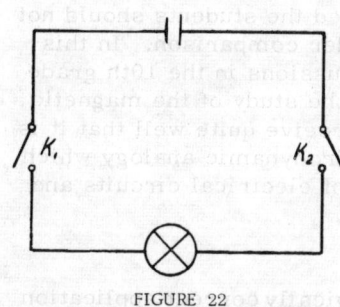

FIGURE 22

With this end in view, an adequate amount of laboratory work was conducted in one of the 7th grades. The students were given to hook up electrical circuits, placing the switch in various parts of the circuit, according to diagrams. After that they had to write a paper, in which a diagram with 2 switches was given (Figure 22), and they were asked whether it was possible to place the switch in either of the two given positions.

In spite of the previously performed laboratory work, most of the students made a mistake and did not give the right answer. Very few of those who gave the right answer could provide an explanation. In view of this, a method of presentation of the problem was worked out, which practically eliminated the error.

It is well known that there are some concepts in the physics course which are formally assimilated; the students learn the definitions by rote, and do not grasp their meaning. The test papers written in class do not always provide the means of detecting the formal elements: the students quite often give clearly formulated answers which have been learned by rote, and which do not actually indicate their true knowledge. Individual discussions with the students are much more useful in such cases.

Discussions of this kind were extremely important in working out a method for the application of analogies in the definitions of the concepts of voltage.

It emerged from the discussions that most of the 7th graders acquire a completely formal conception of the voltage, as the energy of unit charge over a given section of the circuit. The energy itself is for them a complicated and ill-defined concept. The formal way in which the concept of voltage is assimilated was convincingly demonstrated in discussions with the students of the 8th and 9th grades. It turned out that they had very vague notions about the voltage, whereas they were quite clear in their minds about the current, resistance, and output. It is obvious that it is the absence of images in the definition of the voltage in terms of the energy which gives rise to this kind of "knowledge". It is interesting to note that the current intensity is well understood because it is significantly related with the image of a water current, even if hydrodynamic analogies have not been extensively used in the course of teaching. The procedure was subsequently adopted of defining the voltage in the 7th grade both with the help of analogies and in terms of energy, and the teaching results were checked by means of test papers.

The teaching of the 10th-grade students, who have progressed much farther, was also checked by indirect means. For instance, if the students have understood the concept of the potential and not merely memorized its definition, the way of choosing the zero potential should not cause them any trouble. Similarly, if the concept of electrical capacitance is properly understood, the students should not be baffled by the fact that a body can contain any given charge. Wrong answers obviously indicate that these concepts have been poorly understood.

When checking the experimental findings, it was of basic importance to ascertain that if the analogies are properly applied the students should not come to identify with each other the concepts under comparison. In this respect, the best papers and partly also the discussions in the 10th grade have confirmed the fact that when proceeding to the study of the magnetic properties of electrical currents the students perceive quite well that it is inappropriate, and illegitimate, to retain the hydrodynamic analogy which they have extensively used earlier in the study of electrical circuits and some of the laws of current.

It follows from the foregoing that a methodologically correct application of analogy in teaching underscores the common laws governing real processes and phenomena in nature, it enhances the intuitive appeal of teaching and helps students develop their logical reasoning. This methodological procedure facilitates understanding and improves the students' knowledge, and it should find wide application in the methodological literature and in textbooks.

I.M. RUMYANTSEV

THE TRAINING OF STUDENTS FOR PRACTICAL WORK IN PHYSICS

Practical courses in physics were introduced into the curriculum for the 8th to 10th grades of secondary school in the 1954 - 55 school year. They have since become required courses, together with classroom studies.

The purpose of the practical courses is to implement the knowledge the students have acquired in the classroom studies and to train them in independent experimentation. The students have to familiarize themselves with the more complicated technical instruments and equipment in current use; to become acquainted with some of the methods of investigating and determining physical quantities, or establishing laws; and to learn the correct way of interpreting and generalizing the experimental findings. After performing a set of observations, students have to compute the error involved, and learn gradually to deduce independently the range of application of any given measurement, instrument, and observational method.

It sometimes happens, however, that practical courses in physics are not given, mainly due to the lack of suitable equipment and to the fact that teachers have been inadequately trained. These courses are frequently given at the end of the school year, but they are conducted formally and all too often fail to achieve their purpose.

We were able to ascertain this fact when visiting a large number of schools in Moscow and in the Penza and Novosibirsk regions during the inspection tours organized by the Institute of Teaching Methods of the Academy of Pedagogical Sciences of the RSFSR. An appreciable amount of pertinent data was secured by conferring with teachers, by analyzing the answers and test papers of students, and by conducting practical courses with the teachers at the Moscow City Institute for the Advanced Training of Teachers and in the regional courses for professional improvement.

The analysis of the selected data showed that one of the major reasons for the low effectiveness of the practical courses can be attributed to the lack of a systematic preliminary training of the students.

Many teachers make the mistake of treating the practical work separately, as an independent unit. They therefore begin preparatory classes in this subject only at the end of the year, which is not at all a good idea. The training of the students in most cases then begins with a two-hour discussion with the teacher. This usually consists of an introductory talk, intended to inform them of the way in which the laboratory sessions are to be organized, and to acquaint them with the construction and mode of operation of the instruments and facilities which they would have to use and with working procedures. At this point the students also sometimes receive an initial knowledge of experimental methods and the computation of errors.

Given the multiplicity of subjects treated in this introductory discussion the students find it hard to distinguish what is relevant and what is not. They are unable to digest all the information, and thus the talk often fails to achieve its purpose.

Experience has shown that the training of the students for laboratory work should be carried out in a gradual and consistent manner throughout the school year, during the regular classroom studies.

The acquisition of preliminary knowledge bearing directly on future laboratory assignments (e.g., getting acquainted with working methods, with the construction and mode of operation of the apparatus and with the techniques of using it) will focus the students' attention and help rouse their interest in the practical work. To the extent that the students have familiarized themselves with the equipment, they will be better able to use it in practice, and they will accordingly derive more benefits from the laboratory assignments. They will thus become proficient in the practical work, and not just be formally conversant with it.

In the methodological literature particular stress is laid on the preparation of the students for laboratory work. Thus for instance, in his "Methods of Teaching Physics" (Metodika prepodavaniya fiziki), P.A. Znamenskii points out that the physics laboratory work is more successfully performed if the students prepare themselves at home beforehand for each individual assignment. Further on in the book a reference is made to the book by A.A. Pokrovskii and others, "A Physics Laboratory Manual" (Praktikum po fizike), where the training of the students for laboratory work is considered in a broader context and is described as a gradual process, systematically carried out throughout the school year.

However, methods for organizing and carrying out such sustained training have not been adequately worked out as yet, and specific suggestions in this respect are still lacking.

It is known that in the process of teaching, the body of physical ideas of the student must be gradually supplemented and enlarged. Thus for instance in the book by A.A. Pokrovskii and others, "Demonstration Experiments in Physics"(Demonstratsionnye opyty po fizike, Uchpedgiz, 1956), it is stated:

"The set of physical ideas acquired by the students during the observations organized in class and the interest in physics generated thereby permit proceeding in a perfectly natural manner from these ideas to the establishment of the fundamental physical concepts describing elementary phenomena and physical quantities, going on to measurement techniques and the instruments and devices involved, etc.

"These concepts are further elaborated in the other studies (laboratory assignments, problem solution, interrogation) during almost the whole teaching term. The relationships prevailing between them are gradually worked out, thus naturally leading the students to the study of physical laws and theories, i.e., to a sound grasp of the course."

The foregoing applies to a large extent to the need of providing an adequate body of physical knowledge in the course of the year, so as to enable the students to perform their laboratory work successfully.

For instance, having studied in class the quantitative relationship between heat and work, or having investigated the motion of a body thrown horizontally, vertically and at an angle to the horizontal, or the expansion of bodies under the effect of heating, etc., the student should be told how the mechanical equivalent of heat is determined, or shown how the motion of a body is investigated, or taught the method of determination of the linear coefficient of expansion of a solid body using the same instruments that he would have to use in the laboratory.

When doing this it is not only necessary to explain to the student the method of determination of the physical quantities and to demonstrate the instruments, but also to give in brief some general practical instructions as to how to use the instruments; these instructions will be useful further on.

It is obviously not necessary to reproduce in the classroom the whole laboratory procedure, since this would cause an undue loss of time and would not be very convincing, at any rate, as the readings of the measuring instruments would not be visible to the students. In addition, unnecessary details at the preparatory stage may cause the students to lose all interest in the work itself.

What is required is to make the student understand the principle involved, so that he should feel the desire to go on with independent work, to discover a law, to compute a numerical result, to watch the behavior of a phenomenon, etc.

In the course of the preparatory work it thus proves necessary to demonstrate in the classroom the laboratory instruments and facilities and some measurement techniques.

The laboratory instruments are, however, usually designed for individual use. Their readings are accordingly not given to observation by a whole class at the same time. And without being able to display the reading, which is one of the principal purposes of demonstration, there is no point in showing the instruments.

Setting up demonstrations with the laboratory equipment consequently requires special techniques. The teacher should be familiar with them and be proficient in their use.

Various means are available for increasing visibility in experiments displayed in class, but not all of them can be applied in the demonstration of the laboratory instruments. It is thus doubtful that backlighting would improve the visibility of instruments of small dimensions, of their scales and indicators, and of the individual structural details.

There are, on the other hand, various ways of projecting onto a screen the instruments themselves or part of them (shadow, diascopic, episcopic), which can be quite useful for this purpose. When the projection of the instruments themselves is not feasible, use may be made of specially prepared transparencies, drawings, diagrams, photographs, which display the structural details of the instruments or setup.

In some cases it is useful to make simplified demonstration models showing the principle of construction and operation of an instrument, or sometimes to replace the small scales of the instrument by magnified scales and indicators. In this case it is best to accompany the demonstration of the instruments by explanatory drawings on the blackboard.

Let us consider in more detail, by means of some examples, the various methods employed for demonstrating the laboratory instruments within the classroom.

When studying the expansion of bodies under the effect of heating it is necessary to acquaint the student in class with the laboratory instrument used for the determination of the linear coefficient of expansion of solid bodies. After seeing the general aspect of the instrument (Figure 1) and understanding in broad outline its construction, the students have to be shown the movement of the sliding pin and the disc attached to it, which involves one of the main points of the experiment, i.e., how the elongation of a rod is measured.

FIGURE 1

A simple and convenient demonstration method is to make a shadow projection of the part of the instrument which contains the sliding stop and the end of the rod.

To do this the instrument is set on a raised bench between the screen and and the illuminator, projecting on the screen a clear shadowgraph of the fixed support and the end of the tested rod which, on expansion, displaces the disc of the sliding stop (Figures 2a and b). At this point it is also possible to show the student how a micrometer is used to measure the initial distance between the balls sunk into the top part of the disc of the sliding stop and of the fixed support.

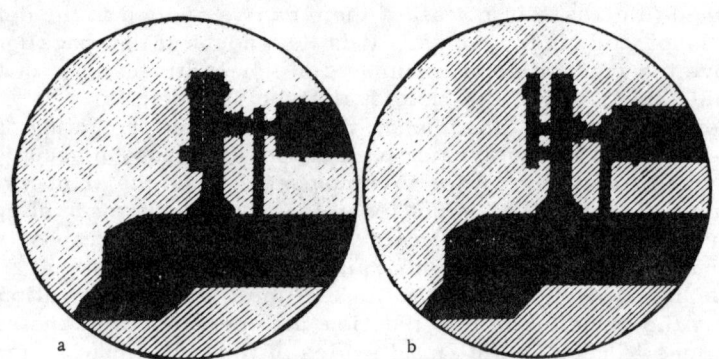

FIGURE 2

In order to avoid heating the body, which would only unduly prolong the demonstration, it is possible to show the increase in the measured distance in the following simple way. It is explained to the students that when the rod is heated it becomes longer, and then the sleeve pipe with the rod is pushed slightly by hand in the direction of elongation and the resultant displacement of the sliding stop is shown. The class will be able to see plainly on the screen the increased distance between the disc of the sliding stop and the fixed support (Figure 2b).

It is then explained to the students that the difference in the two micrometer measurements (before and after heating) yields the value of the elongation experienced by the rod when heated.

Some instruments are more conveniently demonstrated by means of diascopic projection.

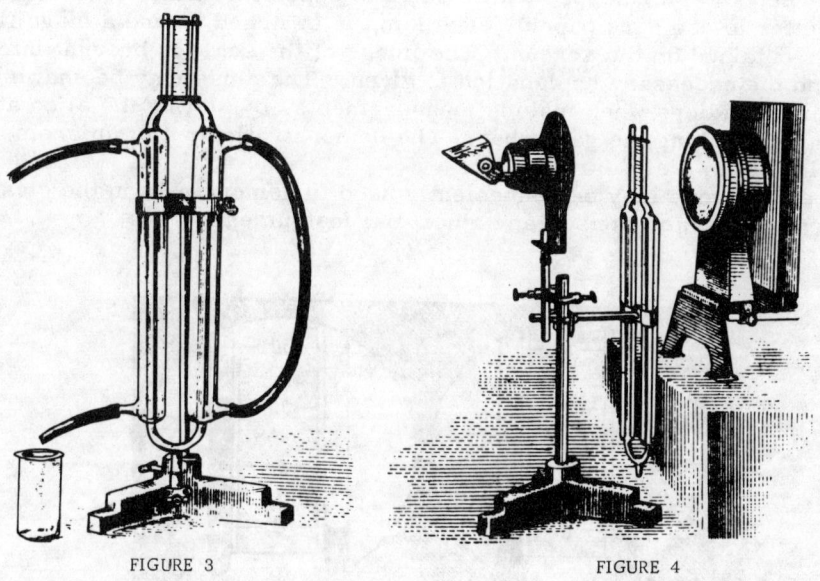

FIGURE 3   FIGURE 4

When the students study the volume expansion of liquids, it is necessary to acquaint them with the laboratory device used for the determination of the volume coefficient of expansion, which consists of communicating vessels (Figure 3). The demonstration involves showing the method of reading the scale of the instrument, which is attached to its upper part.

The setup for projection is assembled as shown in Figure 4. The projector is set on the table, the illuminator housing together with the condensor being shifted forward to the edge of the optical bench. The laboratory device is clamped onto a stand and placed a little below the level of the table, so that its upper part with the scale should be facing the condenser. An objective lens with a reversing prism or a plane mirror is also attached to the stand, and an image is obtained on an inclined screen, as shown in Figure 5.

During the demonstration the student can see on the screen the levels of the tested liquid (kerosene) in the tube: the levels are the same before heating and show a certain difference when the kerosene in one of the arms of the U-tube is heated. The teacher has to explain the correct way of taking the readings when measuring the difference in levels.

There are some cases where it is possible and more convenient to project the instrument or any or its parts on to a screen by means of an episcope.

Thus, when teaching the construction and operation of slide calipers or of a micrometer it is essential to show these instruments and then acquaint the students with the technique of reading the vernier when using these instruments for measurement. Of course, it is possible to make use of specially made, large demonstration models (i.e., large-size vernier gauges for demonstration); this will be discussed below.

This, however, does not exclude the use of episcopic projections, which are quite valuable in their own right, especially since they are very simple to perform. The required instrument is placed on the object stage of an ordinary school epidiascope, with a sheet of paper laid under it (it is sometimes better to use dark paper). The lamp is switched on and a magnified image is obtained on the screen. The image of the scale is brought into focus and the necessary explanation is given. The class may be shown how to perform readings when making measurements, by setting the slide at various points along the gauge bar. The demonstration with a micrometer is carried out in a similar manner.

The epidiascope may be conveniently used to demonstrate in the classroom a pocket stop-watch or any other dial instrument.

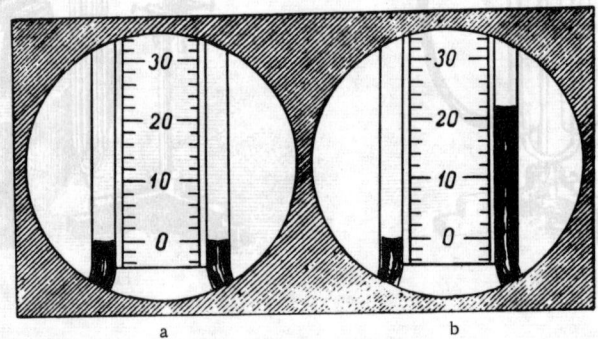

FIGURE 5

Sometimes, as will be shown in the example below, episcopic projection proves to be the only possible means of showing the results of an experiment when demonstrating the laboratory device. Consider the classroom demonstration of a device for the determination of the acceleration of free fall, consisting of a ball and a bar pendulum (Figure 6). The string holding the small metal ball suspended from the top of the bar is burned through, the ball starts falling and at the same time the bar swings back from an inclined to a vertical position. When the bar hits the ball a mark is made on a strip of paper stretched over the ruler, under which there is a length of inked ribbon.

In order to show the marking to the class, the bar is removed from its pivot and the reverse side of the paper strip is projected by means of the epidiascope. The point of impact of the ball will be clearly seen on the screen, giving a measurement of the distance traversed by the falling body, from which the acceleration of free fall can be determined.

In order better to display to the students the construction and operative principle of a given laboratory device involved in measurements, it is possible to make special simplified demonstration models. The problem of making and demonstrating in the classroom such models of various measuring devices has been extensively treated by V. P. Orekhov*.

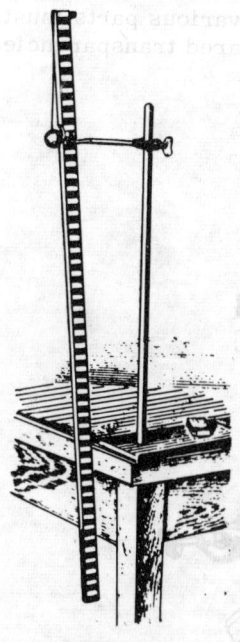

When explaining in the classroom the method for investigating tensile deformations in steel and copper wires on a laboratory device (Figure 7), it is useful to accompany the explanation of the structural features of the device by the demonstration of a simple model made of a rubber band stretched across the blackboard. One end of the rubber band is fixed to the blackboard and the other is passed over a pulley. A load is then applied, for instance by suspending a weight from the rubber band (Figure 8). It is further shown that the test section may be chosen between any two points. Riders are attached at the points selected (the riders may be made of discs of tin foil or cardboard), and their positions are marked on the blackboard with chalk. The students are then shown how the distance between the points increases when the load is increased, while the points are both shifted in the direction of the load.

FIGURE 6

When, during the demonstration, angles have to be read off the scale of the instrument (for instance, when working with a ballistic catapult Figure 9), use may be made of a special scale with a magnified dial, which may be made out of sheet metal or cardboard. Its face is painted white and the divisions are clearly marked on it. This demonstration protractor is attached to the instrument (Figure 10). This simple attachment makes it possible to demonstrate quite clearly many experiments with laboratory instruments and to obtain the necessary numerical data for the solution of problems.

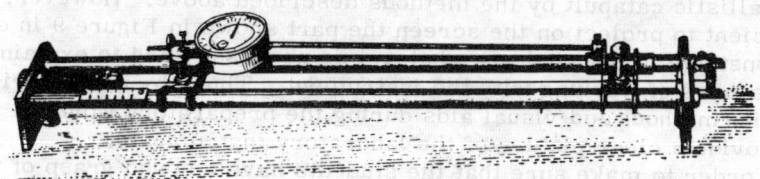

FIGURE 7

Let us give one more example. When demonstrating the device for the determination of the acceleration of free fall (Figure 6), it is necessary to show the students the length of the path traversed by the falling ball up to the

* Orekhov, V. P. Demonstratsii po fizike kak sredstvo podgotovki uchashchikhsya k ovladeniyu izmeritel'-nymi navykami (Demonstrations in Physics as a Means of Training Students in Measuring Techniques). — In: "Uchenye zapiski Ryazanskogo Pedagogicheckogo Instituta," Vol. 14. 1956.

mark made on the bar. To do this, use may be made of a demonstration ruler subdivided into decimeters or centimeters, which have to be laid off on the side of the bar. In the relatively rare cases when the design of the device does not allow the use of the visual aids described above, the explanation of the construction of the device and of its various parts must be accompanied by the demonstration of specially prepared transparencies, drawings, diagrams, or photographs.

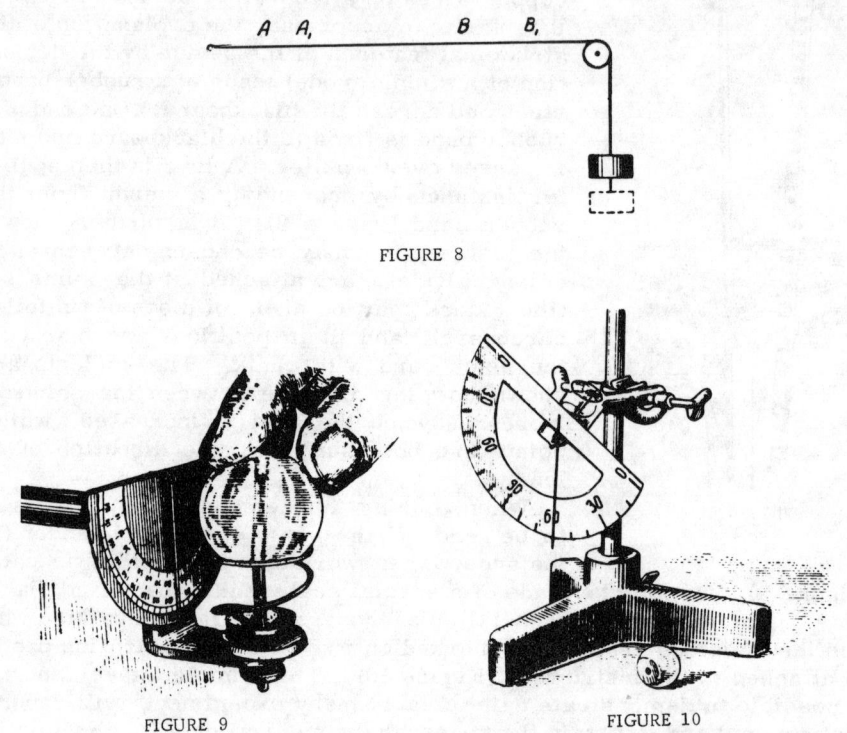

FIGURE 8

FIGURE 9

FIGURE 10

For instance, it does not prove possible to show the structural details of the ballistic catapult by the methods described above. However, it is sufficient to project on the screen the part shown in Figure 9 in order to demonstrate to the students whatever is necessary and to explain the basic procedures of working with the instrument. Thus, by using various demonstration methods and visual aids during the preparatory work, it is possible to provide a clear picture of the laboratory instruments.

In order to make sure that the students have a sound grasp of the preparatory material, it is necessary to review this material when examining the students. They should be able to discuss experimental methods in the same way as they discuss, for instance, Joule's experiment on the determination of the mechanical equivalent of heat. It is useful to have the answers accompanied by a demonstration of the instruments themselves and by a discussion of their construction and of the basic working procedures.

The laboratory instruments give more accurate measurements than do the demonstration devices, which are intended mainly for the study of the qualitative aspect of phenomena. After having provided the necessary visual presentation in his demonstration of experiments with the laboratory instruments, the teacher will therefore have many possibilities of deriving more accurate data for the solution of experimental problems and for checking the results obtained by calculation. It is well known that aritificially devised problems with an abstract content are not associated with concrete images; this detracts from their value in the teaching of physics and reduces the students' interest in their solution. Conversely, problems which are coupled with experimentation always evince in the students a lively interest; they also meet one of the most important pedagogical requirements: that the content of the problems and the numerical data given in them correspond to actual conditions.

In the 8th grade, for instance, after showing in the classroom the device for the determination of the acceleration of free fall (Figure 6), use should be made of it in an experiment to obtain the data necessary for the solution of problems; the data required are the path traversed by the falling body and the time of fall. Given these, one solves the problem of finding the acceleration. Not all the data necessary for the problem have to be obtained in the experiment performed in class. Some of the data may be simply given as a supplement to those derived during the experiment. These data, however, have to be concrete, obtained by the teacher in work with the given instrument when preparing for the lesson. For instance, in the problem involving determination of the acceleration of free fall it is possible to find the path by experiment, and give the time as obtained beforehand.

In the 9th grade, after studying the ballistic catapult (Figure 10), use may be made of it for the solution of problems. Problems such as 143, 154, 167* may thus be "given substance". In these cases the ballistic catapult permits obtaining numerical data and experimentally verifying the results obtained by calculation.

Thus, for instance, problem No. 143 in the collection mentioned above requires finding the velocity with which a missile is shot out of a spring catapult, if when shot vertically up it reaches a height of 110 cms. This problem may be solved with the use of the laboratory device. The ballistic catapult is aimed straight up and the height to which the ball rises is measured.

It is true, though, that under such conditions the experimental nature of the problem is not yet quite apparent, as there is no possibility of verifying the result obtained.

If, however, problem No. 154 is solved next, and the velocity of 4.5 m/sec given in it is replaced by the velocity of the missile as calculated for the given device, it proves possible to check experimentally, by means of the device, the result obtained by calculation. It is understandable that the students should feel satisfaction in finding that the calculated result is identical with the value obtained by experimental means.

Problem No. 167 can also be solved in conjunction with the experimental device. In this case, in order to determine the range of flight of the missile for the angles of 30, 45, and 60°, it is necessary to set in the conditions stated in the problem the shooting velocity as determined for the given device. The calculation results are afterwards experimentally verified.

* Znamenskii, P.A. Sbornik voprosov i zadach po fizike (A Collection of Questions and Problems in Physics). —Uchpedgiz. 1956.

Other experimental problems may also be solved with the use of the ballistic catapult, for instance: 1) determination of the altitude attained by the "missile" when shot at an angle of 45°; 2) determination of the altitude attained by the "missile" when shot straight up; 3) determination of the range of flight in a horizontal shot. Such problems are analyzed in detail in the methodological manuals for teachers.*

The analysis of the errors of measurement and calculation is of importance in many laboratory assignments, as it gives the possibility of evaluating the quality of the work by the numerical results. The familiarization of the students with the computation of errors must also be conducted in a gradual manner. This aspect should be given attention during the preparatory work in class throughout the school year, when solving problems involving actual material derived in given laboratory experiments.

The students should often be asked the question: How would the calculated result be affected if such and such an error is allowed in the measurement of this or that quantity ? The answer to this question will enable the students to see which quantity affects more than others the accuracy of the obtained result.

This will lay the foundation for the selection of the appropriate measuring instruments. The introduction of errors will thus not appear to the student as an abstract matter, but will be intimately linked with the concrete results of problems, experiments, measurements, and ultimately with forthcoming laboratory assignments.

For instance, it will be clear to the student why when determining the mechanical equivalent of heat it is necessary to perform the temperature measurements with an accuracy to within 0.1° and not 1°; why when determining the linear coefficient of expansion of solid bodies, a technical micrometer should be used and the elongation of the rod when heated should be measured with an accuracy to within 0.01 mm; why in the determination of the acceleration of free fall it is enough to measure the path traversed by the freely falling ball in centimeters and not in millimeters.

As is known, the laboratory work is conducted on a different basis from the classroom studies. In the course of classroom studies all the students perform the same laboratory assignment with the same set of instruments, under the direct guidance of the teacher.

During the work in the laboratory each student performs a different assignment and the teacher cannot devote himself to all the students at the same time.

In order that the students should be able to successfully carry out on their own the laboratory assignments, they must be provided with some kind of concise handbook (instruction sheets). The instruction sheet is usually given to the student to take home before conducting the experiment and constitutes one of the stages in the home preparation to the laboratory work.

In reading the instruction sheet the student should get a concrete idea of the concepts formed in the course of the year during the preparatory work in class. For this reason it is of little value to give directions reviewing the known sections of the textbook or answers to questions which do not have a direct bearing on the practical work, as the instruction sheets are often constituted.

---

* Pokrovskii, A.A. et al. Praktikum po fizike (A Physics Laboratory Manual), pp. 70-72. — Uchpedgiz, 1956.

Obviously, the set of instructions cannot be composed in the abstract, without taking into account the given facts and bearing any relationship to the teaching procedure as a whole. Its content must be to a large extent made to correspond with the preliminary work conducted by the teacher during the preparatory classes. Therefore the amount of details contained in the set of instructions will be determined by the coverage given to any particular laboratory assignment beforehand, in the classroom.

For instance, the students will have learned in class how to perform the assignment, the construction of the instrument involved, and some working procedures with it; they will have learned how to prepare the instrument for the experiment and how to perform the experiment. In that case it is enough to mention these things in the instruction sheet briefly, but clearly. A good, clear schematic drawing and a picture of the instrument or set-up together with a concise description of the construction and operation of the device are quite sufficient to help the student in his home preparation and during the performance of the experiment. The instruction sheet should pass in review the necessary material and explain the practical work to be done so that during the laboratory session most of the time should be devoted not to reading the instruction sheet but to performing the experiment, to observation, measurement, processing of the results, and their critical analysis. If the need arises the students should be in a position to modify the experiment or to repeat it.

Let us take one of the laboratory assignments to illustrate the way in which the preparatory work is organized and conducted during the school year. We shall pause to consider the way in which the instruction sheet for this assignment is composed. The example considered will be the 9th-grade laboratory assignment "Determination of the mechanical equivalent of heat."

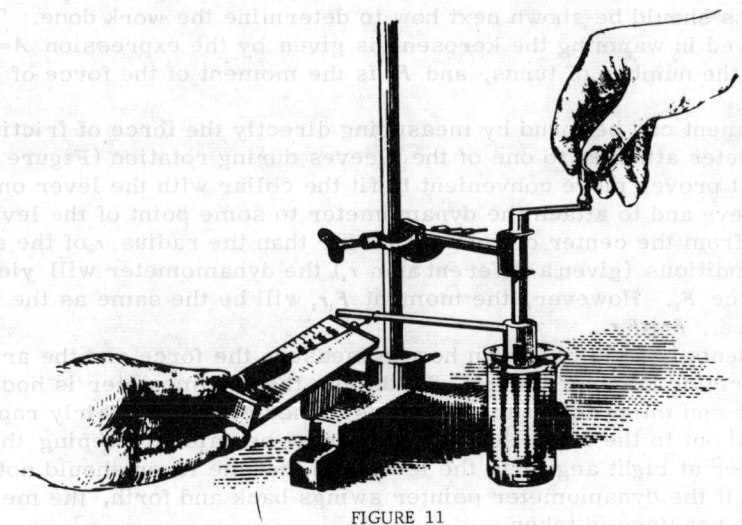

FIGURE 11

The experiment is performed with a device of new design, of which a general view is given in Figure 11. The components of the setup and the procedure for assembly are described further on. The students are shown in class the construction of the device, they are required to examine its basic components, and they are shown how the setup is assembled for the experiment. The demonstration of the device should be accompanied by projection of the components by means of an epidiascope and by the appropriate drawings on the blackboard. It is further necessary to show the ways of measuring and calculating the work of the force of friction and the amount of heat evolved.

There is no need to perform the whole experiment in class as this would take too much time, and furthermore the change in temperature will not be visible to the whole class. However, the procedure to be followed in performing the experiment should be thoroughly explained to the students.

When the crank of the device is turned, the lever connected to the outside sleeve is pushed against the stand and prevents the sleeve from turning. The inner sleeve keeps turning at the same time, and as a result the sleeves warm up because of the friction between them. The amount of heat evolved is measured by the usual calorimetric method. To do this the sleeves are fitted one inside the other and plunged into a beaker with kerosene. The class is shown how this is done. The change in the temperature is measured with a thermometer, to within 0.1°.

Given the mass of the sleeves, the kerosene, and the beaker, the specific heat of brass, kerosene, and glass, and the initial and final temperatures, the amount of heat evolved can be calculated. At this point it is helpful to show the students how to remove the sleeves from the device in order to weigh them. The outer sleeve is held with the left hand and the crank is turned counter-clockwise until the sleeves, together with the collar piece, become unscrewed from it. The collar piece is then twisted clockwise so that the slots in it face the lugs on the outer sleeve, and the collar is pulled off. It is pointed out to the students that the sleeves can be weighed fitted together since there is no need to take them apart during the experiment.

The class should be shown next how to determine the work done. The work involved in warming the kerosene is given by the expression $A = 2\pi r F n$ where n is the number of turns, and Fr is the moment of the force of friction.

This moment can be found by measuring directly the force of friction with a dynamometer attached to one of the sleeves during rotation (Figure 12). However, it proves more convenient to fit the collar with the lever onto the outside sleeve and to attach the dynamometer to some point of the lever, at a distance from the center of rotation larger than the radius r of the sleeve. In these conditions (given a different arm r_1) the dynamometer will yield another force F_1. However, the moment $F_1 r_1$ will be the same as the first moment, i.e., $Fr = F_1 r_1$.

The students have to be shown how to measure the force and the arm when performing the experiment. To do this the dynamometer is hooked on to the lever and the crank of the device is turned evenly but fairly rapidly. It is pointed out to the class that attention must be paid to keeping the dynamometer at right angles to the lever and that the lever should not touch the stand. If the dynamometer pointer swings back and forth, the mean value of the readings is taken.

The arm is determined by measuring with a ruler the distance along the lever between the point where the dynamometer is connected and the line drawn across the center of the collar.

It should be explained to the students that this way of measuring the force, when the dynamometer is attached to the lever rather than to the crank and holds the outside sleeve, makes it possible to take into account only the force of friction whose work goes into heat and eliminates from consideration any additional forces of friction in the other components of the device.

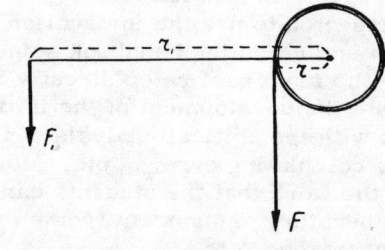

FIGURE 12

Finally it is necessary to calculate the mechanical equivalent of heat. For this purpose the students are given a problem. The conditions of the problem must involve actual quantities, obtained from experiment with the device. The teacher can obtain these quantities by performing the experiment when preparing for the lesson.

Let the mass of the sleeves be m_1 = 28.7 g, the mass of the kerosene m_2 = 80 g, the mass of the glass beaker m_3 = 29 g, the initial temperature t_1 = 20.9°, the final temperature t_2 = 22.3°, the force F = 160 g, the radius r = 10 cm, and the number of turns n = 300.

Determine the mechanical equivalent of heat.

The work $A = Fs$, or $A = F \cdot 2\pi r n = 0.16 \cdot 2 \cdot 3.14 \cdot 0.1 \cdot 300 = 30.14$ kgm.

The amount of heat $Q = (m_1 \cdot c_1 + m_2 c_2 + m_3 \cdot c_3) \cdot (t_2° - t_1°)$ or $Q = (0.093 \cdot 28.7 + 0.51 \cdot 80 + 0.2 \cdot 29) \cdot (22.3° - 20.9°) = 69$ cal $= 0.069$ kcal. The mechanical equivalent of heat $I = \frac{A}{Q}$, or $I = \frac{30.14}{0.069} = 437$ kgm/kcal.

Thus, the students have been shown in class how to perform the experiment, and they should know it and be able to discuss it. In addition, the students have been acquainted with the construction of the device, have received some practical instructions how to use it, have taken the necessary notes and made the necessary drawings in their notebooks, and have carried out the calculations.

All this has to be taken into account when compiling the instruction sheets for the students. Thus, for instance, during the preparation in class it has been explained to the students in some detail how to calculate the work performed on heating the kerosene. A schematic drawing (Figure 12) has been used to explain the equality of the moments $Fr = F_1 r_1$, and it has been shown how to measure the force with a dynamometer connected to the outer sleeve when turning the crank. These explanations may therefore be dropped from the instruction sheets. But as the students have to recall the material pertaining to this particular section, the sheet must contain an appropriate set of questions. The students must prepare answers to them on the basis of the explanations received in class, with the help of the notes they took and the drawings they made.

The object of the experiment is also clear to the students from the work in class, but it should be briefly considered in the handbook. Also a brief but clear account should be given of the construction of the instrument and of the basic working procedures with it. Thus, the set of instructions for the students should cover: the object of the experiment, particulars about the construction of the instruments, the working procedure to be followed, the method of noting down results and calculations.

Since the students are expected to use the instruction sheets when preparing at home, it is helpful to include in the handbook some control questions. These questions should for the most part refer directly to the experiments to be performed and promote the development of the initiative and independence of the students, help with the critical analysis of the obtained results and with the correct way of calculating errors, etc. Moreover, the questions should refer to the part of the work that the students can do at home during their preparation for the laboratory assignment (some preliminary observations, measurements, comparisons, etc.).

We give as an example one of the instruction sheets for the students, composed with due allowance of the preparatory work done in class.

Assignment No. 2. DETERMINATION OF THE MECHANICAL EQUIVALENT OF HEAT BY THE CALORIMETRIC METHOD

Equipment (Figure 13): (1) Device for the determination of the mechanical equivalent of heat ; (2) a 200-500 g technical arm balance with a sensitivity of 0.01 g; (3) a set of weights of 200 g (weights ranging from 0.01 g to 100 g); (4) a thermometer for 0° to 30° (or 50°), divided into 0.1°-0.2°; (5) a laboratory dynamometer for 400 g, divided into 10 g; (6) a 30 cm ruler, divided into millimeters; (7) a 100 cc chemical beaker, (8) a flask of kerosene; (9) a stand with a coupling joint.

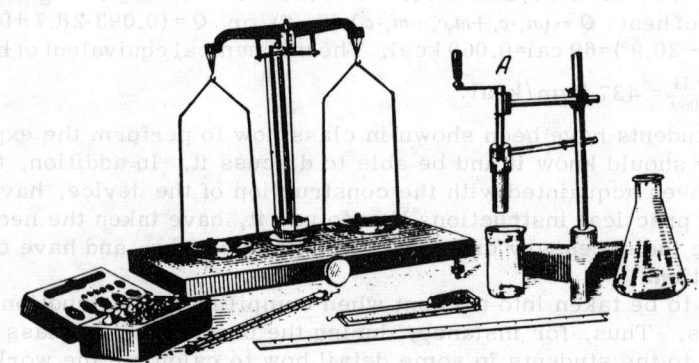

FIGURE 13

Object of the experiment

The mechanical equivalent of heat establishes the quantitative relationship between heat and work. It denotes how many units of work are equivalent to a unit quantity of heat.

It has been shown on the basis of many theoretical and experimental investigations that the ratio of the work A performed to the amount of heat Q derived therefrom, i.e., $J = \frac{A}{Q}$, is a constant magnitude, equal to 472 kgm/kcal.

The device for the determination of the mechanical equivalent of heat (Figure 14) consists of two brass sleeves (1) and (2) fitted one into the other; a collar piece with a lever (3) for measuring the force of friction and the arm; a crank (4) for turning the inner sleeve; a bracket with an extension rod for attaching the device to the stand. The inner sleeve (2) is cut along spiral lines down its length and the resultant helicoidal strips stick slightly out so that when the sleeve is inserted into the outer one it is firmly pressed against its wall.

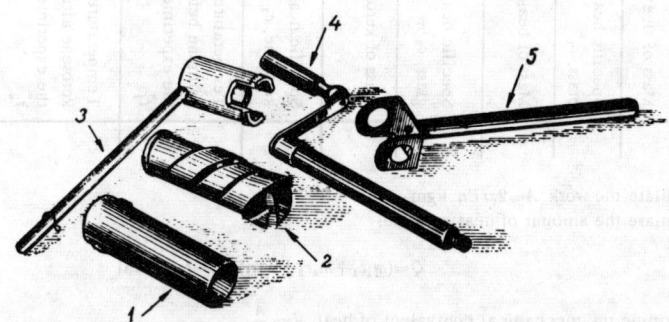

FIGURE 14

The device is assembled for the experiment in the following manner. The sleeves are fitted one into the other and the collar with the lever is slipped on top so that the slots in the collar receive the lugs protruding from the outer sleeve; the crank is then slipped into the bracket with the extension rod and the spindle is screwed into the center hole of the inner sleeve. The whole device is fastened to the coupling joint on the stand, so that when the crank is turned clockwise the lever comes to rest against the column of the stand and thus check the rotation of the outer sleeve.

When performing the experiment the sleeves of the assembled device are placed into a beaker with kerosene and the crank is turned. As a result of friction the sleeves warm up and cause the kerosene in the beaker to warm as well. Given the mass of the sleeves, the kerosene, and the beaker, as well as the specific heat of brass, kerosene, and glass, after determining the initial and final temperatures it is possible to calculate the amount of heat Q evolved.

The work done to warm up the kerosene is given by the expression $A = 2\pi rFn$, where n is the number of turns, F is the force as measured by the dynamometer attached to the lever which keeps the outer sleeve from rotating, as shown in Figure 11, r is the distance along the lever from the point where the dynamometer is attached to the axis of rotation.

Will the moment Fr of the force of friction be the same if the dynamometer is attached at other points on the lever (farther or closer to the axis of rotation)? What is the disadvantage of taking the lever too long or too short in the measurements?

Execution

1. Weigh the chemical beaker, first empty and then with the kerosene. Pour in about 70-80 g of kerosene. Calculate the mass of the kerosene.

2. Weigh the sleeves, one inserted into the other, after first removing them from the device. To do this, hold the outer sleeve with one hand, unscrew the crank and take off the collar piece with the lever.

3. Assemble the device for the experiment and turn the crank until the lever comes to rest against the column of the stand. On rotating the crank further the inner sleeve should keep turning with it and rub against the inside of the outer sleeve, which is held fixed. Put the beaker with kerosene beneath the sleeves and lower the coupling joint of the stand until the sleeves are completely immersed in the kerosene.

4. Mix the kerosene thoroughly and measure its initial temperature, $t_1°$, as accurately as possible.

5. Turn the crank fairly rapidly but evenly, and count the number of turns (for instance, up to 300). Mix well the kerosene again, and measure the final temperature, $t_2°$, to within one tenth of a degree.

6. Hook the dynamometer onto the lever and turn the crank evenly, holding the dynamometer so as to keep the lever from coming into contact with the stand. Be careful to hold the dynamometer horizontal and at right angles to the lever. Note down the dynamometer reading. If the pointer of the dynamometer wavers, take the mean of the readings.

7. Measure the lever arm r, i.e., the distance between the dynamometer hook and the center of rotation.

Record the measurement results in a table:

Mass of sleeves, m_1	Specific heat of brass, c_1	Mass of beaker, m_2	Specific heat of glass, c_2	Mass of kerosene, m_3	Specific heat of kerosene, c_3	Temperature of kerosene before the experiment, $t_1°$	Temperature of kerosene after the experiment, $t_2°$	Force F	Lever arm r	Number of turns, n

Calculate the work $A = 2\pi r F n$ kgm.
Calculate the amount of heat evolved:

$$Q = (m_1 c_1 + m_2 c_2 + m_3 c_3) \cdot (t_2 - t_1) \text{ kcal}$$

Determine the mechanical equivalent of heat $J = \dfrac{A}{Q}$.

Questions. 1. Does the mechanical equivalent of heat you have obtained agree with the value given above? Which measured quantity affects most the accuracy of the result?

2. How will the experimental result be affected by an error of 0.1° in the measurement of the temperature? By an error of 5 g in the force?

3. How many joules of work are equivalent to one calorie?

4. What is the value of the thermal equivalent of work, which denotes the amount of heat corresponding to one kgm of work? To one joule of work?

Experience has shown that it is better to make the set of instructions in the form of detached sheets rather than a bound handbook.

Each instruction sheet must contain the necessary drawings and pictures, be well printed, and clipped to a stiff cover the size of a student notebook with the number and title of the assignment on it. It is helpful to make the cover of each individual assignment of a different color. This will facilitate the distribution and exchange of the instruction sheets among the students.

In order to supply a whole class (thirty students) with instruction sheets, five different assignments can be prepared, each in a batch of six units, to make up a set of thirty. A number of such sets corresponding to the number of parallel classes should be kept in the physics laboratory, to be distributed among the students for home study and exchanged after the performance of the experiments.

The preparatory training of the students for laboratory work in the course of the year, with the help of suitably graded equipment and the instruction sheets drawn up for this purpose, will prove of material assistance at the end of the year when the laboratory work is performed.

In that case the introductory session before the laboratory work can be made to give general information. At that point the students are already acquainted with the assignments, they are familiar with the instruments involved, and they have been given some practical directions how to carry out the experiments; it is thus sufficient just to mention these in the introductory session. On the other hand it is necessary to explain clearly the procedure to be followed in the course of the laboratory work, to show how written reports are prepared, and to emphasize the importance of thoroughly preparing at home in order to know how to perform the experiment successfully and to handle the equipment properly; it should be pointed out that perfect order must be maintained on the working premises, that the experiments have to be independently performed, and the preliminary drafts of the reports should be made. It is necessary to make clear the safety rules to be observed when performing experiments with electrical power sources and with heating units.

In the 8th grade, where the students are confronted with this kind of work for the first time, part of the time in the introductory session must be given to the study of measurement techniques and the analysis of errors. Detailed material on these questions can be found in methodological reference books.*

It should be explained to the students in the introductory session how to write a report of the performed experiment. One or two laboratory assignments already known to the students can be taken to illustrate the way in which the reports are prepared. The report should reflect the students' knowledge of the subject and show how this knowledge has been applied to the solution of the concrete experimental problem. It is helpful to produce samples of reports from previous years.

It has been found that the report should contain: a schematic drawing of the set up with which the experiment was performed, specification of the basic purpose of the experiment, the procedure followed, the results of observations and measurements, the processing of the results, i.e., the calculation of errors, the plotting of graphs, etc. There is obviously no need to copy out of the instruction sheet particulars about the instruments and materials or remarks on the way in which they are used, i.e., whatever is not directly relevant to the knowledge and proficiency of the student.

The experimental work conducted in the 1957-58 school year at the Moscow schools Nos. 204 and 315 showed that the successful performance of the laboratory assignments and the proper organization of the work of the students are to a large extent determined by the quality of the preliminary training in class and of the instruction sheets provided for the experiments, and by the preparation of the students at home.

If these prerequisites are satisfied, the teacher can organize his time most efficiently during the laboratory work.

The teacher can first of all easily ascertain how well the students are prepared for the assignments. A simple observation of the fact that some of the students procede promptly and confidently with the assembly of the setup and the performance of the experiment, while others spend the time reading the instruction sheet which they have not studied at home, indicates that the latter are inadequately prepared and that they require special attention.

The teacher can next watch the amount of care that different students devote to the experiment: some apply themselves and strive to secure the best possible results, others perform the work superficially and formally, without stopping to think about it.

These observations can be quite indicative and, in conjunction with the written reports submitted by the students, they enable the teacher to assess the progress made by each student in the laboratory work.

This overall appraisal of the students' level of knowledge, which is based not only on the written reports but on the quality of the work as a whole,

---

* Pokrovskii, A. A. et al. Praktikum po fizike (A Physics Laboratory Manual). —Uchpedgiz, 1956; Znamenskii, P. A. Laboratornye zanyatiya po fizike v srednei shkole (Laboratory Work in Secondary School Physics), Part 1. —Uchpedgiz, 1955; Kamenetskii, S. E. Vychislenie i otsenka pogreshnostei v laboratornykh rabotakh po fizike v srednei shkole (The Calculation and Estimation of Errors in Secondary School Physics Laboratory Assignments).—In: sbornik "Iz praktiki politekhnicheskogo obucheniya." APN RSFSR, Moskva, 1955.

fosters a responsible attitude in the students and discourages the practice of merely trying to write a good report. The students thus learn to prepare seriously for the assignments and to devote the proper care to the performance of the work.

The preparatory training, together with a properly organized method of conducting the laboratory work and a well-chosen set of equipment, ensures the high level of success expected from the laboratory work. This has been confirmed by the experimental test carried out in the 1957-58 school year.

V.G. RAZUMOVSKII

TECHNICAL CREATIVITY OF STUDENTS IN PHYSICS HOBBY GROUPS

The "Law on the Consolidation of the Link between the School and Life and the Further Development of the Public Educational System in the USSR" has confronted the school with the task of training the students for creative work in production. Thus the objective of the Soviet school is not only to impart to the students a knowledge of scientific principles, to inculcate the necessary skills and habits, and to familiarize them with the basic modes of production, but also to develop their abilities for creatively applying the acquired knowledge and skills.

This task is intimately connected with the problem of developing the ability for creative work, which has for a long time been an object of study of leading psychologists and pedagogues.

The Soviet materialistic psychology espouses the viewpoint that the only innate characteristics underlying the development of abilities are those of a physio-anatomic nature; the abilities themselves are always the result of development, which is mainly realized in the course of education and training. Abilities can be manifested and developed in no other way than within the framework of a given concrete activity.

In order to develop creative abilities in the technical domain, it is obviously necessary to engage in systematic activity within that domain.

Any technical occupation is principally based on physics. It is therefore most logical to link the organization of the technical creativity of youngsters with the teaching of physics. As the experience of the Soviet school has shown, this link not only ensures the development of creative abilities but it also promotes a sound grasp of the theoretical knowledge.

A particularly favorable time to do this is in the senior classes. At that age the students begin to form lasting interests, the creative imagination becomes very active, and the accumulated knowledge becomes systematized.

The experience of many teachers, especially in the post-war years, has shown that extracurricular work is most conducive to the development of the creative abilities of students. This refers primarily to physico-technical hobbies.

However, the method of conducting extracurricular activities and developing the creative abilities of students in physico-technical hobby groups has not yet been adequately developed.

The methodologies of I. I. Sokolov and P. A. Znamenskii dealing with the work of hobby groups give suggestions that are mainly of an organizational nature, such as: the hobby group is formed on a voluntary basis; the group should be composed of students from different classes (to provide continuity in the work); the projects of the students are prepared under the scientific guidance of the teacher — he stimulates the general interest, supervises the preparation of reports, and gives his conclusions on the performance of the students.

Nothing is said in these methodologies as to the concrete way of organizing the group activities and of developing the creative abilities of the students.

It is true that Professor Znamenskii's methodology makes the significant methodological suggestion that "when the student is designing and building a device his attention should be constantly drawn to the physical principle involved in the construction and operation of the individual parts..." This recommendation ("to draw the attention to the physical principle involved") does not, however, resolve the problem of how to apply creatively the theoretical knowledge in practice, and how to develop the proficiency for technical work. "The object of consciousness is not that to which attention is directed, but rather that which is the purpose of the action performed, i.e., its direct outcome".*

Errors stemming from a lack of understanding of this fact are quite common in the methodological literature on technical hobbies for the young.

For instance, in "Technical Hobbies" (Tekhnicheskoe tvorchestvo, Molodaya gvardiya Press, 1955), the section "The physico-technical hobby group" describes how to build some instruments. It might appear at first sight that after building a model of the pulley balance described on p. 130 of the book the student would get a better idea of the moment of a force. Such is not the case, however. The observations we made at the Central Station of Junior Technicians showed that the 8th graders who built this device by the drawing and the description given in the book made no use at all of the concept of the moment of a force they had studied in school. This is obviously due to the fact that the description, which merely gives specifications and a drawing for building the device, does not call for any independent thought or action involving the determination of the moment of a force.

The description for a homemade ampere-voltmeter of the moving-iron type, which is given on p. 131, is a typical example of a technical prescription restricting to the utmost the possibility of displaying any creativity (at least as far as one's knowledge of electromagnetism is concerned).

The description of the rotary wind-driven generator for powering radio installations (p. 143) is quite valuable for the physico-technical hobby groups. But here again the instructions given are of the kind that induce the student only to perform manual operations and do not incite him to make conscious use of his knowledge in physics. It is actually quite easy to explain to the students on what the output of the wind generator depends and

---

* Leont'ev, A. N. Psikhologicheskie voprosy soznatel'nosti ucheniya (Psychological Aspects of Consciousness of Learning). — Izvestiya APN RSFSR", No. 7. 1957.

how it is calculated. Other books for young hobbyists also suffer from the same deficiency, for instance the book "Do it Yourself" (Svoimi rukami, Trudrezervizdat, 1957). This is on the whole quite a good and useful collection for junior amateur technicians, but it does not solve the specific problem of stimulating inventiveness and imaginative thinking.

It is an important task for the physics teacher to channel the activities of his group into the practical application of their knowledge and proficiency. This will ultimately determine the interest of the students towards the subject and the extent to which their knowledge of physics will be enhanced.

The question as to what constitutes the object of activity of the junior technician in each individual case is a fundamental methodological problem.

In order to make this point clear, let us refer to the experimental work we conducted at the N. M. Shvernik's Central Station of Junior Technicians (TsSYuT). The group members, who were 8th graders, were divided into two groups, "A" and "B". Both groups were given the same assignment: to build a model of the pulley balance described on p. 130 in the book "Tekhnicheskoe tvorchestvo" (Technical Hobbies).

"The device consists of a wooden pulley 300 mm in diameter, with a hole 50 mm in diameter sawn out near its edge and filled with lead.

"The pulley is grooved over its circumference, and a length of twine is run in the groove. One end of the twine is firmly attached to the pulley, and the tray of a balance is suspended from the other end. The pulley is screwed, through a hole in its center, onto a vertical wooden support, so that it should rotate freely. The support is nailed to a wooden base measuring 200 × 200 mm. After the balance is ready, it is graduated by means of calibrated weights."

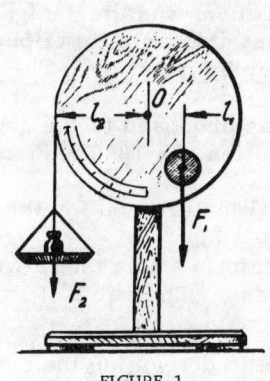

FIGURE 1

The students are first reminded of what the moment of a force is, and of the condition of equilibrium for the forces applied to a body having an axis of rotation. A diagram is drawn on the blackboard, to explain the physical principle involved in weighing something on the pulley balance (Figure 1). The system is in equilibrium when the moment of the gravitational force of the weighed body is equal to the moment of the gravitational force of the counterweight, i. e., when $F_2 l_2 = F_1 l_1$.

Group "A" is then told to use the description given in the book "Technical Hobbies" and build the device.

In order to speed up the building of the device, each member of the group was given a different assignment (to make a particular piece, or to perform a given operation). In this way the device was ready by the end of the second session.

An identical assignment was carried out by group "B". But when the lead counterweight for the device was made and its mass was determined (120 g), the students were told to make a preliminary calculation: what is the maximum weight that can be weighed on this balance. This additoinal assignment required little extra time, and the members of group "B" were gratified to find that the experimental test yielded a value agreeing with the expected one. It turned out that weights up to 100 g could be measured on the finished balance.

In the following session the members of both groups were assigned a control work: to design a balance for weighing bodies up to 200 grams. Most of the members of group B managed the task well, and made sketches and calculations in their notebooks. Of group A only one half came out well. The designs proposed by the students of group B also showed more variety: some changed the size of the counterweight, others the diameter of the pulley, and yet others changed both of these. On the other hand, almost all the projects in group A involved a counterweight of the same mass and exhibited a dependence on the old design.

This work, and individual talks with the students of the two groups, thus showed that the members of group B had a much better grasp of the concept of the moment of a force and of the conditions of equilibrium for the forces applied on a body having an axis of rotation. It is manifest that the object of awareness is not determined either by the subject or by "drawing the attention" of the student to it. What is actively recognized is whatever is involved in the activity, as the direct object of a given practical action.

Therefore, in order to channel the technical creativity of the students into the application of their school knowledge of physics, it is necessary to guide their activity in the appropriate manner.

At the Central Station of Junior Technicians, in the wind-power group for 8th and 9th graders, the following procedure was adopted. The members of the group were required to build a series of wind installations of increasing complexity. The main task in each of them was the design of a specified item. The problem could be solved with the knowledge of physics from school, or with the help of some supplementary information. The rest was simplified in one way or another.

For instance, the following practical problem was proposed to the group.

There is a bicycle dynamo having a speed of rotation $n = 1800$ rpm and an output $N = 10$ W.

It is required to design a wind-powered electrical installation for the illumination of a point.

The instructor of the group must first of all explain in a talk that a wind motor utilizes the kinetic energy of a moving air mass. The value of the energy may be found by the formula $W_k = \frac{mv^2}{2}$, which is known to eighth graders. On the basis of this formula it is possible to determine the output of the air stream flowing across the area circumscribed by the rotor of the wind motor. If R is the radius of the wind rotor, the volume of air flowing through during one second will be $V = \pi R^2 v$, its mass $m = V\rho = \pi R^2 v \rho$, and therefore the power $N = \frac{\pi R^2 v^3 \rho}{2}$.

But the wind motor utilizes only part of the energy possessed by the wind stream. The number denoting the fraction of the energy of the wind stream utilized by the wind motor is called the power coefficient ξ. The output of the wind motor is therefore defined by

$$N_1 = \frac{\pi R^2 v^3 \rho}{2} \xi.$$

This yields the radius of the wind rotor for a given output, i.e.,

$$R = \sqrt{\frac{2N_1}{\pi v^3 \rho \xi}}.$$

In places where the mean yearly wind velocity is low it is necessary to use wind rotors with many vanes.

A second important question arises: What should be the transmission ratio from the wind rotor to the wheel of the dynamo?

The students are given a definition of the speed of the rotor, as the ratio between the velocity of the tip of a vane and the wind velocity, i.e., the tip speed ratio $Z = \frac{\omega R}{v}$, where R is the radius of the rotor, v is the wind velocity, and ω is the angular velocity of the wind rotor.

Given a wind rotor with 18 vanes, we find from the appropriate tables cf., E. M. Fateev's book: Wind Motors and Wind-Power Installations (Vetrodvigateli i vetroustanovki, Mashgiz. 1956) that the tip speed ratio $Z = 0.9$, i.e., the peripheral velocity of such a wind rotor will be about the same as the wind velocity (3 to 8 meters per second), and this is precisely the peripheral velocity of a normally running bicycle wheel (the velocity of a bicycle rider). In the present case, therefore, the best thing to do is to make a friction transmission directly from the rim of the wind rotor.

The design and dimensions of the other components are given to the group by the instructor. The participants are moreover relieved from making the more intricate parts and they are provided with all the facilities for completing the model within the shortest possible time. When engaged in manual work the students concentrate only on the procedures and operations that are being studied at the given moment (i.e., in the 8th and 9th grades).

FIGURE 2   FIGURE 3   FIGURE 4

A view of the low-speed wind-powered electrical installation made by the group members at the N. M. Shvernik TsSYuT is shown in Figure 2.

After the model has been tested the group is required to make a high-speed installation of the same output.

A third problem arises: Find the maximum radius of a high-speed wind rotor ($Z=7$) which rotates at a rate of 1800 rpm at a wind velocity of 8 m/sec.

The relationship $Z = \frac{\omega R}{v} = \frac{2\pi n R}{v}$ shows that when the radius increases the speed of rotation decreases (since Z is a constant for a given type of rotor). From $Z = \frac{\omega R}{v}$ we obtain $R = \frac{Zv}{\omega} = \frac{Z_1 v}{2\pi \cdot n}$.

Substituting the numerical values, we get

$$R = \frac{7 \cdot 8}{2 \cdot 3.14 \cdot 30} = 0.3 \text{ m}.$$

Let us calculate whether a wind rotor (with $\xi = 0.4$) having a radius of 0.25 m will have a sufficient output (fourth problem). Now

$$N = \frac{\pi R^2 v^3 \rho \cdot \xi}{2}; \quad N = \frac{3.14 \cdot 8 \cdot 8 \cdot 8 \cdot 0.12 \cdot 0.4}{16.2} = 0.76 \text{ kgm/sec} = 7.44 \text{ W},$$

i.e., this wind rotor has a sufficient output to run a bicycle dynamo (Figure 3).

If a more powerful wind-powered electrical installation is made, moderators of the speed of rotation have to be put on the vanes of the rotor (to prevent the rotor from flying apart in extreme cases). One way of making such brakes is shown in Figure 4.

When the wind rotor is running at a normal speed the brake flaps are in the position ab, so that the air flows smoothly past them as the rotor turns. They are held in this position by a special spring fixed in the blade of the vane.

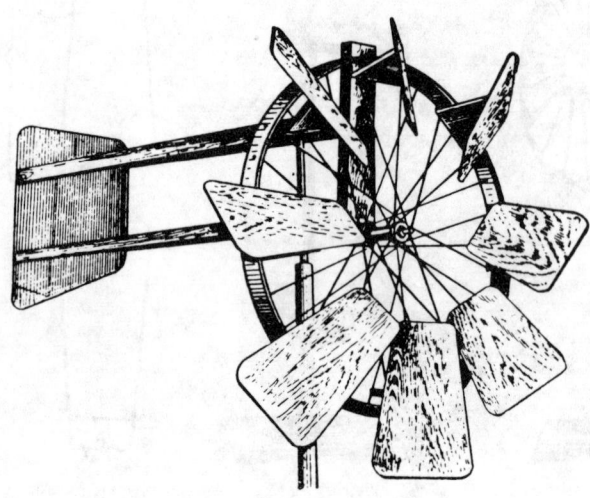

FIGURE 5

Each flap has a flyweight attached to its end, which exerts a centrifugal force $F=\frac{mv^2}{R}$ that increases with the speed of rotation. Determining the tension of the spring (by means of a dynamometer), it is possible to calculate the mass of the flyweight which would move the flaps into the position $a'\,b'$ when a certain speed of rotation is reached (fifth problem). Thus

$$m=\frac{FR}{v^2}=\frac{FR}{4\pi^2 R^2 n^2}=\frac{F}{4\pi^2 R n^2},$$

where F is the tension of the spring (on the arm of the flyweight), R is the distance of the regulator from the axis of the rotor, and n is the speed of rotation (number of revolutions) at which the air brake is brought into operation.

This way of organizing the work in the physico-technical hobby groups provides the means of applying creatively the knowledge acquired in the physics classes. Further, this method of organization develops in the junior technicians the proper approach to building design, proceeding from the principle of operation, the principal parameters, and the constituent elements of the designed structure.

The fact that physical problems are formulated in concrete terms (i.e., the requirement to build a model) enhances the students' interest in their knowledge, on which they have to draw, and increases their proficiency in it.

An analysis of the work performed over one year by one of the members of the group (the eighth grader Vladimir G. of Moscow School, No. 204) showed that at the beginning of his activities his work was mostly imitative, but at the end of the year it showed signs of true creativity. Thus, for instance, this student designed and constructed a model of a wind motor, using for the purpose an old bicycle wheel (Figure 5). Similar progress was also made by other members (Zhenya S., Lyuba G., Ira P., and others). As a result these students achieved a sounder and more effective knowledge of physics.

QC
30
A4353

APR 27 1970